T0387908

Behavior and Design of Trapezoidally Corrugated Web Girders for Bridge Construction

Woodhead Publishing Series in
Civil and Structural Engineering

Behavior and Design of Trapezoidally Corrugated Web Girders for Bridge Construction

Recent Advances

Authored by

Mostafa Fahmi Hassanein

School of Civil Engineering and Geomatics,
Southwest Petroleum University, Chengdu, Sichuan, China;
Department of Structural Engineering, Faculty of Engineering,
Tanta University, Tanta, Gharbia, Egypt

YongBo Shao

School of Civil Engineering and Geomatics,
Southwest Petroleum University, Chengdu, Sichuan, China

Man Zhou

School of Civil Engineering, Wuhan University, Wuhan, Hubei, China

WOODHEAD
PUBLISHING
An imprint of Elsevier
elsevier.com/books-and-journals

ELSEVIER

Woodhead Publishing is an imprint of Elsevier
50 Hampshire Street, 5th Floor, Cambridge, MA 02139, United States
The Boulevard, Langford Lane, Kidlington, OX5 1GB, United Kingdom

Notices
Knowledge and best practice in this field are constantly changing. As new research and experience broaden our understanding, changes in research methods, professional practices, or medical treatment may become necessary.

Practitioners and researchers must always rely on their own experience and knowledge in evaluating and using any information, methods, compounds, or experiments described herein. In using such information or methods they should be mindful of their own safety and the safety of others, including parties for whom they have a professional responsibility.

To the fullest extent of the law, neither the Publisher nor the authors, contributors, or editors, assume any liability for any injury and/or damage to persons or property as a matter of products liability, negligence or otherwise, or from any use or operation of any methods, products, instructions, or ideas contained in the material herein.

ISBN: 978-0-323-88437-2

For Information on all Woodhead Publishing publications visit our website at https://www.elsevier.com/books-and-journals

Publisher: Matthew Deans
Acquisitions Editor: Glyn Jones
Editorial Project Manager: Fernanda A. Oliveira
Production Project Manager: Debasish Ghosh
Cover Designer: Christian J. Bilbow

Typeset by Aptara, New Delhi, India

Working together to grow libraries in developing countries

www.elsevier.com • www.bookaid.org

Contents

Introduction

1

1.1 General

Corrugated web girders (CWGs) are those girders with webs composing of a series of longitudinal and inclined folds. Because CWGs have many advantages over I-girders with transversely-stiffened flat webs (IPGs), they have been extensively used as bridge girders globally. For example, there is no interaction between shear and flexure taking place in CWGs thanks to the minor axial stiffness of the corrugated webs (CWs). This is often called in the literature as "the accordion effect". Accordingly, prestressed concrete hybrid bridges require less prestressing compared to conventional prestressed concrete bridges. With respect to their fatigue behavior, CWGs have been found to own higher fatigue strengths, compared to IPGs, mainly because they do not utilize welded transversal stiffeners. Overall, using CWGs is currently spreading so fast due to their superior resistance to shear and out-of-plane stiffness. Since each fold in a CW is supported by adjacent folds, the stiffness in the out-of-plane direction of CW is greater than that of the flat web. Based on the above advantages, several investigations have focused on the response of CWGs to shear and flexural loadings.

On the other hand, the grades of structural steels are becoming much higher than before thanks to the great improvement recently taken place in the production process of steel materials accompanied by the need to design lighter structures. Hence, the term "high strength steel (HSS)" becomes familiar nowadays in the construction sector. HSSs are considered as those steels having a minimum yield/proof stress (F_y) of $460MPa$. Recently, such materials have progressively been utilized in several engineering structures, especially as heavily loaded structural members because of their high strength-to-weight ratio. HSS provides weight savings, hence it results in reduced fabrication, transportation, and erection costs in modern construction. Moreover, in comparison with conventional steels of normal strength (NSSs), the application of HSS reduces the (1) overall consumption of the material and (2) release of carbon dioxide in society. For global development, steel is utilized in large masses in structural applications and with the fast growth of towns, with the New Capital of Egypt and Chengdu city of China are just examples, the need for structural steels becomes greater. However, with steel production, the energy consumptions and gas emissions are the major environmental concerns. According to World steel association (WSA), applying HSSs instead of NSSs reduces the steel material total release of 0.156 billion tons CO_2 equivalents. China, the native country of most of this research group, was globally the highest country in steel production in 2015 by 803.8 million tons. This highlights the importance of replacing NSSs with HSSs in modern constructions in order to save the environment by efficiently decreasing the carbon dioxide discharges from steel production process. Accordingly, the literature shows that a considerable research has recently been directed to investigate steel elements formed from S460 and S690. Based on the above, several researches have been developed to combine the advantages of the HSSs and the CWGs to provide economical girders.

Behavior and Design of Trapezoidally Corrugated Web Girders for Bridge Construction: Recent Advances.
DOI: https://doi.org/10.1016/B978-0-323-88437-2.00001-0

Additionally, the advances in structural and fabrication technologies have also led to the development of the hollow tubular flange girders, with the main aim of increasing the lateral-torsional buckling of these girders. However, a design engineer should ensure that the designed element is safe under different possible failure modes, from which the shear failure being one of them. This encouraged the current authors to investigate the effect of adding these tubular flanges to I-girders by suggesting using the hollow tubular flange plate girders rather than the conventional I-plate girders (IPGs). To the authors' best knowledge, there exist at least two practical engineering instances of bridge girders with tubular flange girders filled with concrete and CWs. The first one is the Maupre Bridge, built in france, which consists of a box girder of triangular section, a concrete-filled tubular bottom flange and a prestressed concrete deck. The other one is the 30 m span prestressed composite bridge with bottom tubular flange girders filled with concrete, which is built in China. This new type of composite structure consists of superimposed concrete slab with steel plate, corrugated steel webs, and rounded-ended rectangular tubular bottom flange filled with concrete. The excellent mechanical properties have been verified by a full-scale experiment.

Based on the above introduction, it seems that research on CWGs with different configurations and materials has been expanded to include many variables. The current research group, indeed, has contributed to the development of such elements, as will be seen in this book which aims to provide the designers and researchers with the recent development done in CWGs.

1.2   Objectives

The object of this book is to deepen the understanding of the behavior of CWGs used in bridges. Thus, the main objective is to develop different design procedures for the CWGs under different straining actions and materials used. Furthermore, the goals are extended to study the effect of web-to-flange juncture and recent development of erection methods. Also, the effect of using tubular flanges on the behavior of the CWGs is additionally considered.

1.3   Book organization

This book contains an introduction besides *eight* chapters. Chapter *one* is concerned with the introduction of the book. Chapter *two* provides the development of bridges with CWs. Chapter *three* focuses on the real boundary condition between flange and web of CWGs. Chapter *four* provides the shear buckling behavior depending on the results provided by the authors. Chapter *five* describes the flexural buckling behavior of the CWGs. In chapter *six*, the stress analysis of I-girders with concrete-filled tubular flange and CW is made. Chapter *seven* is providing the recent erection methods of the bridges formed with CWGs. Finally, chapter *eight* concludes the book by providing the conclusions, the recommendations, and the further studies to be investigated.

Development of bridges with corrugated webs

2

2.1 General

Prestressed concrete bridges with corrugated steel webs (CSWs) are composite structures first developed in France [1]. In recent years, this structural type has begun to attract the attention of Chinese engineers. In China, the first prestressed concrete bridge with CSWs (Long-March Footbridge) was completed in 2005 [2]. To date, more than 40 bridges of this structural type have been constructed, some of which are detailed in Table 2.1. Indeed, China ranks second in the world in the use of prestressed concrete bridges with CSWs. With the accumulation of engineering experience and advances in design technology, the hybrid prestressed concrete bridge with CSWs has developed to cover long spans and complex types. For example, Zhuhai Qianshanhe River Bridge, which is a three-span continuous box girder bridge with 160 m in maximum span, is the longest bridge in the world in the same type of bridges [3]. The main navigational bridge of the Chaoyang Bridge in Nanchang, which is a composite box girder cable-stayed bridge with CSWs, is composed of six pylons and the stay cables are arranged in a single cable plane [4]. After more than 15 years of development, China has made tremendous progress in the theoretical research, design, and construction of this new type of structure. Many scholars from various universities and institutes in China have made extensive research on the basic mechanical properties (bending, shear, and torsion behavior), shear and torsional buckling, analysis of the shear connector, and the dynamical characteristics [5–13]. To expand the knowledge of this type of bridges worldwide, this paper summarizes the characteristics of prestressed concrete bridges with CSWs, and describes their basic properties, typical construction examples, and the new construction techniques in China.

2.2 Mechanical feature

Fig. 2.1 shows the basic components forming a prestressed concrete bridge with CSWs. In this bridge type, the concrete webs of a conventional prestressed concrete bridge, which account for 20%–40% of the dead load, are replaced by CSWs. In addition to the substantial reduction in the dead load of the main girder, this system provides excellent structural characteristics due to the excellent mechanical properties of the CSWs. Folded steel webs have a high resistance to shear forces, while their resistance to axial forces or bending moments is almost negligible owing to the so-called "accordion effect" of CSWs. Accordingly, their axial rigidity is much smaller than that of the concrete slabs, so the prestressing force can act efficiently on the concrete flanges. The application of CSWs in prestressed concrete bridges is structurally rational because

Behavior and Design of Trapezoidally Corrugated Web Girders for Bridge Construction: Recent Advances.
DOI: https://doi.org/10.1016/B978-0-323-88437-2.00008-3

Table 2.1 Details of prestressed concrete bridges with corrugated steel webs built in China in between 2005 and 2017.

No.	Bridge name	Bridge type	Bridge span (m)	Years of completion
1	Huai'an Changzheng bridge	Continuous box girder bridge	18.5 + 30 + 18.5	2005
2	Guangshan Pohe river bridge	Simply supported-to-continuous box girder bridge	4 × 30	2005
3	Dayanhe river bridge	Simply supported box girder bridge	25	2005
4	Ningbo Yongxinhe river bridge	Continuous box girder bridge	24 + 40 + 24	2006
5	Dongying Yingzuo B bridge	Simply supported box girder bridge	38	2007
6	He'nan Weihe river bridge	Continuous box girder bridge	47 + 52 + 47	2010
7	Juancheng Yellow river bridge	Continuous box girder bridge	70 + 11 × 120 + 70	2011
8	Xin'mi Zhenshuihe river bridge	Cable stayed bridge without backstays	30 + 70 + 30	2011
9	Guangzhou Yuwotou bridge	Continuous box girder bridge	35 + 50 + 35	2012
10	Taohuayu Yellow river bridge	Continuous box girder bridge	75 + 135 + 75	2013
11	Xingtai Nanshuibeidiao bridge	Continuous box girder bridge	70 + 120 + 70	2013
12	Neimenggu Jingjiawan bridge	Continuous box girder bridge	44 + 3 × 80 + 44	2013
13	Nanchang Chaoyang bridge	Cable stayed bridge	79 + 5 × 150 + 79	2015
14	Changzhuang Ganqu bridge	Continuous box girder bridge	9 × 50 + 9 × 50 + 40	2016
15	Qianshanhe river bridge	Continuous box girder bridge	90 + 160 + 90	2016
16	Chaoyanggou bridge	Extradosed bridge	58 + 118 + 188 + 108	2017

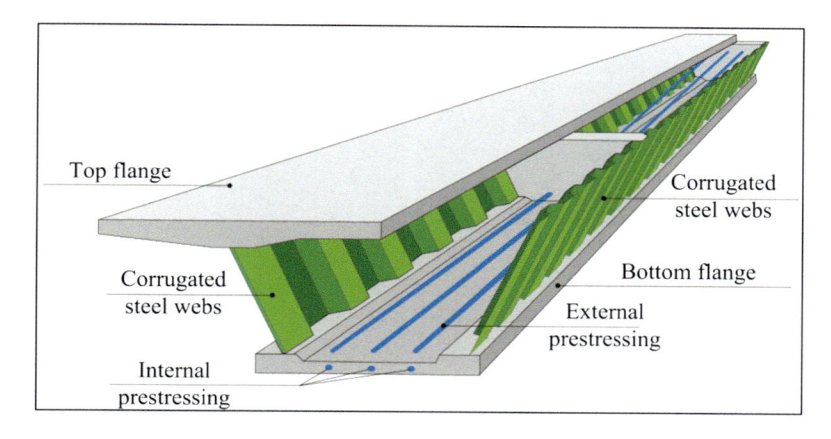

Figure 2.1 Prestressed concrete box girder with CSWs. *CSWs*, corrugated steel webs.

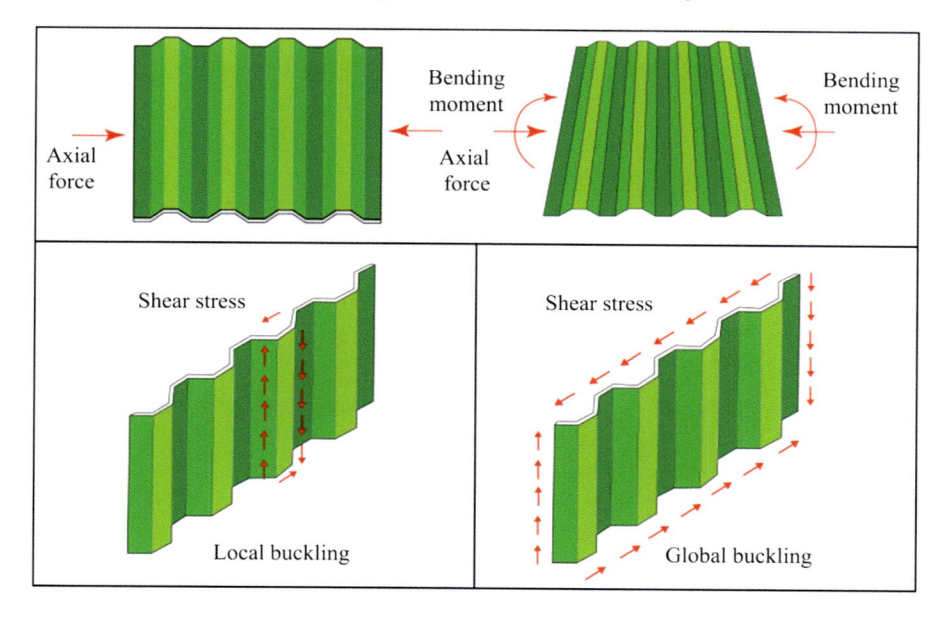

Figure 2.2 Properties of CSWs. *CSWs*, corrugated steel webs.

the prestressing efficiency is increased and the shear forces are adequately resisted. Further, the working effort is reduced, since formworks and reinforcement for the concrete webs are eliminated.

2.2.1 Bending behavior

As shown in Fig. 2.2, due to the accordion effect of box girders with trapezoidal CSWs, the axial rigidity of a CSW is much smaller than that of a concrete slab and it can be neglected from the engineering viewpoint. Thus, it is widely accepted that the bending

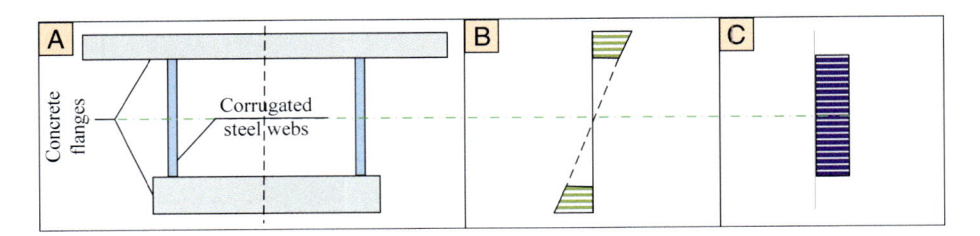

Figure 2.3 Stress distribution of box girder with CSWs: (A) cross-section; (B) the distribution of the normal stress; (C) the distribution of the shear stress. *CSWs*, corrugated steel webs.

moments are entirely carried by the concrete flanges, whereas the webs alone carry the shear forces which are distributed evenly over the web height [14–15]. Additionally, the quasi-plane assumption is valid in this kind of structures, that is, the concrete flange plane section in a girder bridge with CSWs remains nearly flat after bending [8].

2.2.2 Shear behavior

Trapezoidal CSWs used in box girder, as shown in Fig. 2.2, can provide improved shear buckling strength compared to flat steel webs so that stiffeners can be saved and steel consumption can be reduced. Due to the accordion effect discussed above, the uniform shear stress distribution over the web height (Fig. 2.3C) allows determination of the shear strength of the girder without considering the moment–shear interaction. Additionally, the shear strength of trapezoidal CSWs may significantly influence the ultimate shear loading capacity of the whole box girder [16].

2.2.3 Torsion behavior

Due to the weakening of the cross-section, the torsional rigidity of box girder with CSWs is about 40% of that of conventional box girders with concrete webs [17]. However, the torsional mechanism of this composite structure is more complicated because both the concrete flanges and the CSWs provide resistance to the externally applied torque. Note that the concrete flanges can continue to carry the torque even after the buckling (or yield) of the CSWs. Therefore, as shown in Fig. 2.4, a closed shear flow in the top and bottom concrete flanges will be developed, and the shear flow in the composite cross-section will no longer be continuous. Accordingly, the shear flow in the CSWs will not be equal to that in the concrete flanges. Generally, the ultimate torsional capacity of the composite section can be obtained by the fixed angle softened truss model [18].

2.2.4 Buckling behavior

Corrugated webs have been found to fail in shear by instability (i.e. web shear buckling). Three types of buckling of the CSWs under shear forces have been studied, namely, the local buckling, global buckling, and interactive buckling, as shown in Fig. 2.5. In theory, local buckling involves a single flat panel or "fold," global buckling

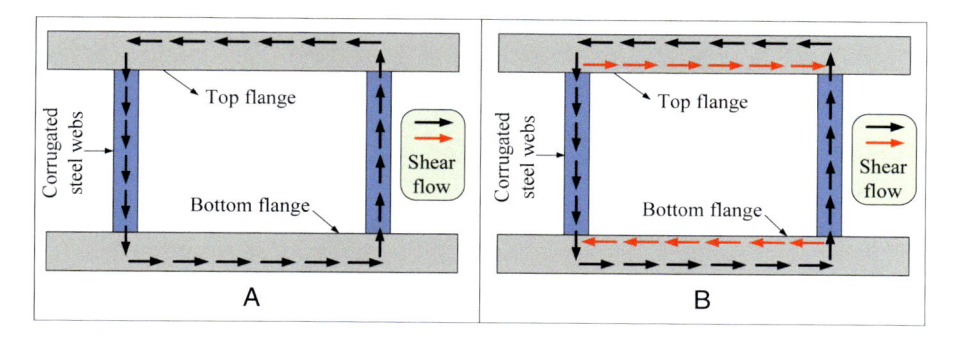

Figure 2.4 Distribution of shear stress on cross-section: (A) before CSWs' buckling or yielding; (B) after CSWs' buckling or yielding. *CSWs*, corrugated steel webs.

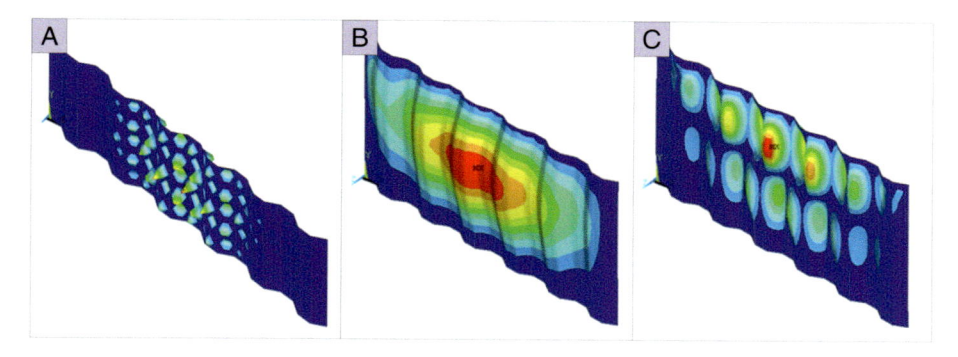

Figure 2.5 Shear buckling modes of CSWs: (A) local buckling; (B) global buckling; (C) interactive buckling. *CSW*, corrugated steel web.

involves multiple folds, with buckles that extend diagonally over the entire depth of the web, whereas interactive buckling involves a single flat panel and multiple folds simultaneously. Since CSWs cannot be expected to provide any post-buckling strength, it has been recommended to design this type of bridges by confirming that the webs do not buckle as long as the ultimate state has been reached [19–20].

2.3 Structural layout

2.3.1 Joint between corrugated steel webs and concrete flanges

The joint between the CSW and the concrete slab is an important part of this type of bridges, from which the shearing force along the longitudinal direction is transferred and the stripping between the concrete flanges and the CSWs can be prevented [21–24]. With the development of this steel-concrete composite structure, multiple shear connectors have been developed to pursue the best overall mechanical performance. As shown in Fig. 2.6, ordinary joint structures are classified according to the types of the shear connectors.

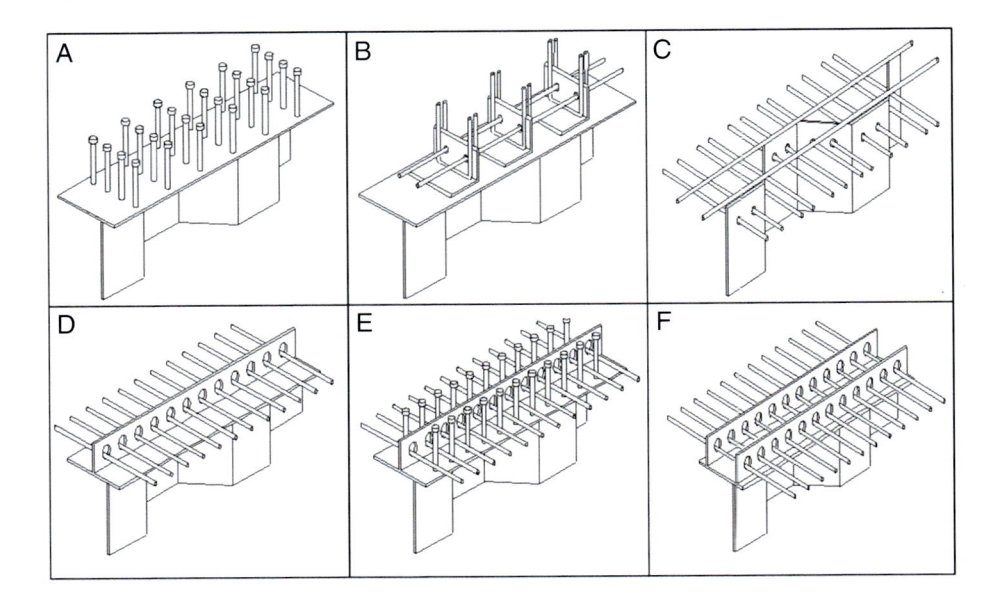

Figure 2.6 Types of joints between CSW and concrete slab: (A) stud connection; (B) angle connection;(C) embedded connection; (D) S-PBL connection; (E) S-PBL + stud connection; (F) T-PBL connection. *CSW*, corrugated steel web.

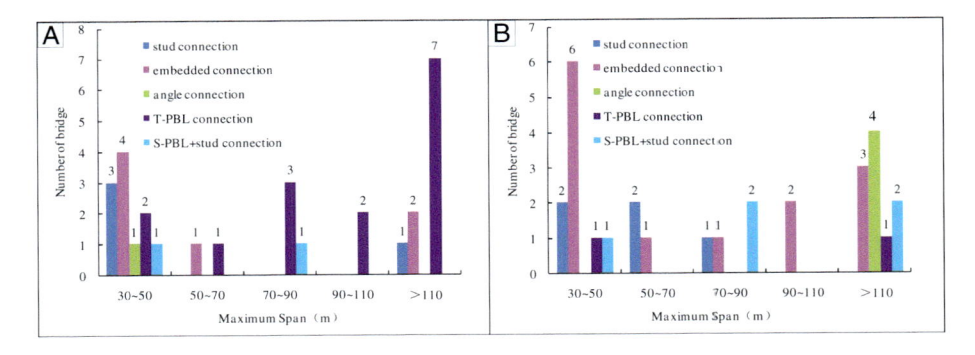

Figure 2.7 Types of joint between CSW and concrete slab: (A) connection between corrugated webs and supper concrete slab; (B) connection between corrugated webs and lower connection slab. *CSW*, corrugated steel web.

Fig. 2.7 provides statistics for the types of joints between the CSW and the concrete slab for 29 prestressed concrete bridges with CSWs constructed in China. As shown in the chart, stud connection and embedded connection are both widely used to connect the concrete slabs (including the upper and lower slabs) and the CSW in small span bridges with CSWs. With span increasing, the joint between the CSWs and the upper concrete slab mainly adopts the T-PBL connection, while it mainly utilizes the embedded connection and the angle connection with the lower concrete

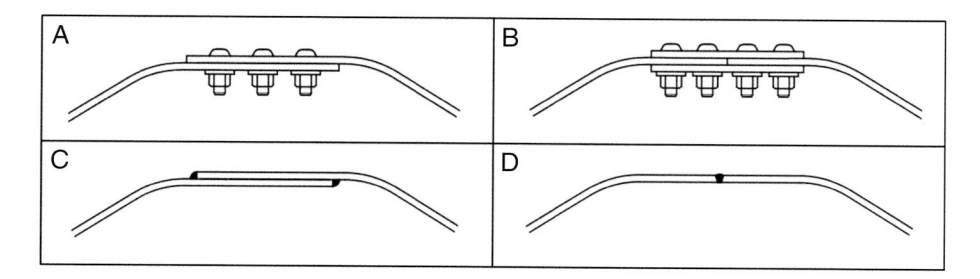

Figure 2.8 Types of joint between CSWs: (A) single shear friction joint; (B) double shear friction joint; (C) lapped fillet welded joint; (D) butt-welded joint. *CSWs*, corrugated steel webs.

slab. Because there is no need for flange plates to allow fitting of shear connectors, it is simple to solve the errors during erection and economic. T-PBL connection and angle connection are more rigid than the stud connection and they are relatively inexpensive because of their simplicity of welding.

2.3.2 Joint between corrugated steel webs

Due to the processing length limit, CSWs must be connected with each other. Since there are no axial forces acting on the CSWs of a prestressed concrete bridge, there is no need to align the axes of the webs. Accordingly, single shear friction joint and lap fillet welded joint can be used. These joints provide a much simpler structure compared with ordinary double shear friction joint and butt welded joint, thus, the work efforts and costs are effectively reduced. The most common connection forms between CSWs are shown in Fig. 2.8. It is suggested that butt welded joint is to be adopted to make the longitudinal connection of adjoining CSWs and the horizontal connection when the CSWs need to be joined above each other within the depth of the web.

2.3.3 Corrugated steel webs with concrete encasement

It is generally known that shear bucking is a key issue in the design of girder bridges with CSWs, as shown earlier. Accordingly, to improve the structural buckling strength in long-span bridges, the CSWs near the intermediate supports are usually encased with concrete within a specific distance from the bridge pier. Fig. 2.9 shows a schematic of this composite web.

2.4 Typical construction applications

(1) Long-March Footbridge (Fig. 2.10A):
  The Long-March Footbridge is the first prestressed concrete bridge with CSWs in China, which was constructed in 2005. It is a (18.5 + 30.0 + 18.5) m continuous girder bridge with single-cell box, 7.0 m in width and 1.6 m in height. In order to meet the requirements of navigation during construction, the construction technology of full space support has been used in the bridge construction. It was erected using temporary pier in the mid-span [2].
(2) Juancheng Yellow River Highway Bridge (Fig. 2.10B):

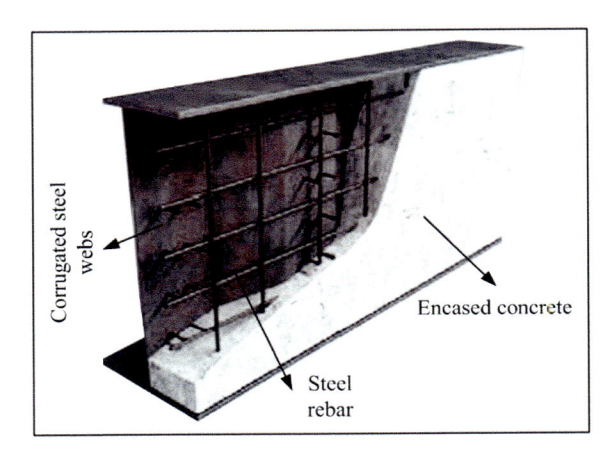

Figure 2.9 Layout of CSWs with concrete encasement. *CSWs*, corrugated steel webs.

Figure 2.10 Typical application of bridges with CSWs: (A) Long-March Footbridge; (B) Juancheng Yellow River Highway Bridge; (C) Chuhe River Bridge; (D) Qianshanhe River Bridge; (E) Nanchang Chaoyang Bridge; (F) Chaoyanggou Bridge. *CSWs*, corrugated steel webs.

The Juancheng Yellow River Highway Bridge is a 13-span continuous box girder bridge with a total length of 1460 m, a maximum span of 120 m, and a width of 13.5 m. It was the world's longest prestressed continuous girder concrete bridge with CSWs. The construction technique of Juancheng Yellow River Highway Bridge was the cantilever construction. A specially made truss was used for positioning and installation of the CSW. The CSWs were firstly temporarily fixed by bolts and then connected by field welding [25].

(3) Chuhe River Bridge (Fig. 2.10C):
The Chuhe River Bridge is the first large-span non-prismatic continuous girder bridge with CSWs in China, from which its span arrangement is (53 + 96 + 53) m. The main girders consist of a single-cell box cross-section with a width of 16.55 m. They were erected by the free balanced cantilever casting construction method. To solve the problem of positioning the spatial locations of the webs during the construction of the bridge, the mathematic

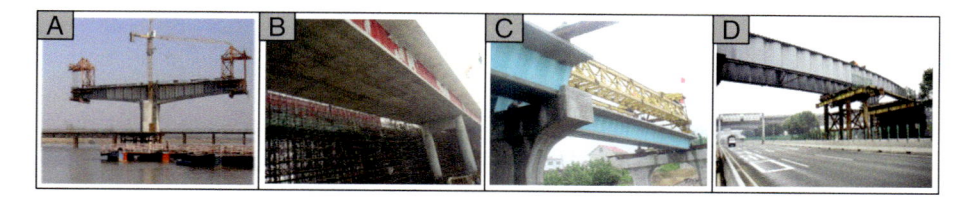

Figure 2.11 Conventional construction methods of bridges with CSWs: (A) cantilever construction; (B) full frame construction; (C) precast-assemble construction; (D) non-brace porued construction. *CSWs*, corrugated steel webs.

theorem was considered to exactly locate the bridge in the plan. By following this theorem, the location was determined when three points were spatially not on a straight line [26].

(4) Qianshanhe River Bridge (Fig. 2.10D):

The Qianshanhe River Bridge is a three-span continuous box girder bridge with main girders consisting of a single-box cross-section. It is 340 m in overall length, 160 m in maximum span, and 15.75 m in width, 9.5 m in height at the pier top and 4.0 m at mid-span. Accordingly, this bridge is the world's largest span continuous box girder bridge with CSWs. Its main piers' section is a uniform rectangular solid pier and the pier caps are separated between the right and left lines. Bored filling pile, with a diameter of 2.8 m, has been used as the pile foundation. It was constructed by the cantilever erection method [3].

(5) Nanchang Chaoyang Bridge (Fig. 2.10E):

The Nanchang Chaoyang Bridge, which is a $(79 + 5 \times 150 + 79)$ m continuous girder cable-stayed bridge, consists of a five-box cross-section with a totla length of 1610 m, a maximum span of 150 m, and a width of 43.8 m. The bridge is the world's first multi-pylon cable-stayed bridge with CSWs and a double deck. The pylon looks like the Chinese character "合," which signifies the blessing of good fortune. The cantilever construction method was applied to erect this bridge [4].

(6) Chaoyanggou Bridge (Fig. 2.10F):

The Chaoyanggou Bridge is a four-span composite box girder extradosed bridge with main girders. It is composed of a four-cell box cross-section. It is 484.8 m in overall length, 188 m in maximum span, and 35.0 m in width. It is the largest span extradosed bridge in the world and it was constructed by the cantilever erection method [27].

2.5 New construction technologies

With the improvement of technology in the erection field of the prestressed concrete bridge with CSWs in China, the construction technology is getting richer and more mature. Common construction methods for prestressed concrete bridge' with CSWs' are similar to those used with ordinary prestressed concrete girder bridge, such as cantilever construction, full-frame construction, precast-assemble construction, and non-brace poured in situ construction, as shown in Fig. 2.11 [28,29]. With advances in construction technologies, new construction methods include rapid cantilever construction of ripple web (RW cantilever construction), which uses CSW as a load bearing component (Fenghua River bridge in Ningbo city and Toudao River bridge in Chengdu

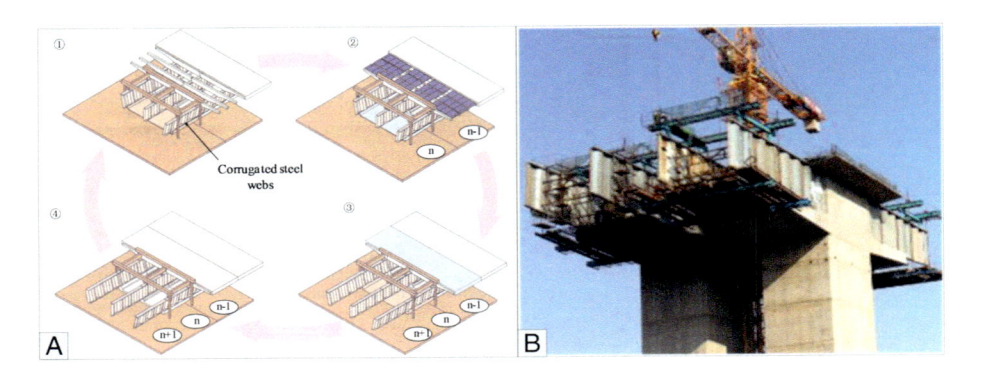

Figure 2.12 RW cantilever construction: (A) construction process and (B) RW cantilever construction stage. *RW*, ripple web.

city, China). The incremental launching construction that utilizes the CSW as a guiding beam (Changzhuang Reservoir viaduct in Zhengzhou city, China) also provides a good construction method for prestressed concrete bridge with CSWs.

2.5.1   RW cantilever construction

For the large-span bridge with CSWs, the traditional cantilever construction with diamond-shaped basket has the following disadvantages: (1) the production cost is relatively high because the basket has a larger amount of steel, and the dead weight of the basket causes a significant negative impact on the girder; (2) the structure of the traditional diamond-shaped hanging basket is complex and it is difficult to control the deformation of the basket during the construction; (3) the lifting space of the CSWs is limited by the height of the hanging basket which increases the difficulty during construction and thus affects the construction period.

RW cantilever construction (Fig. 2.12) is a new construction technology for composite bridge with CSWs. This method uses CSWs to cover the weight of the hanging basket weight and to improve the efficiency of the material. The hanging basket of RW cantilever construction is simplified and weighs much less than the traditional diamond-shaped hanging basket. In additional, the construction platforms of the top slab, the bottom slab, and corrugated steel webs are staggered which may greatly improve the construction efficiency. It is, therefore, much better than the traditional cantilever construction with the diamond-shaped basket [30].

2.5.2   Incremental launching construction

Using the incremental launching method in the construction of bridges with CSWs offers the advantage of leaving the space under the girders unoccupied. Thus, this construction method is not affected by the terrain and buildings adjacent to the construction site. Therefore, it is an effective method for constructing bridges when

Figure 2.13 Incremental launching construction.

there would be difficulties in installing a supporting structure, such as bridges across existing railroads and roads, bridges with high piers and bridges in mountainous regions [31,32].

A steel guide girder is required for a concrete bridge which is rarely applied in other bridge engineering. Conversely, for a bridge with CSWs, the webs can be used as guide girder to save construction cost effectively. It is possible to eliminate the special prestressing steel, usually required for launching purposes, and use a rational arrangement of prestressing steel for the completed structural system if CSWs with continuous upper and lower flanges in the bridge axis direction are launched in advance. By reducing the dead load during launching, this method also leads to a simplified launching system. After completion of the launching work, the lower slab and transverse girders are constructed. The main girder cross-section is completed by constructing the upper slab, and finally the required prestressing is introduced by using the external tendons. Construction of the lower slab is streamlined by placing precast prestressed concrete panels in advance on the lower flange to be used as scaffolding and formwork. Accordingly, this method has recently been adopted for prestressed concrete bridges with CSWs, as shown in Fig. 2.13.

2.6 Future developmental trends of bridge with corrugated steel webs

This new-style composite bridge with CSWs is widely applied in China, which ranks second globally in the number of constructed bridges (just behind Japan). Fig. 2.14 shows a comparison of the spans of the bridges with CSWs between China and Japan, from which it can be noticed that this bridge type is mainly applied in Japan in medium

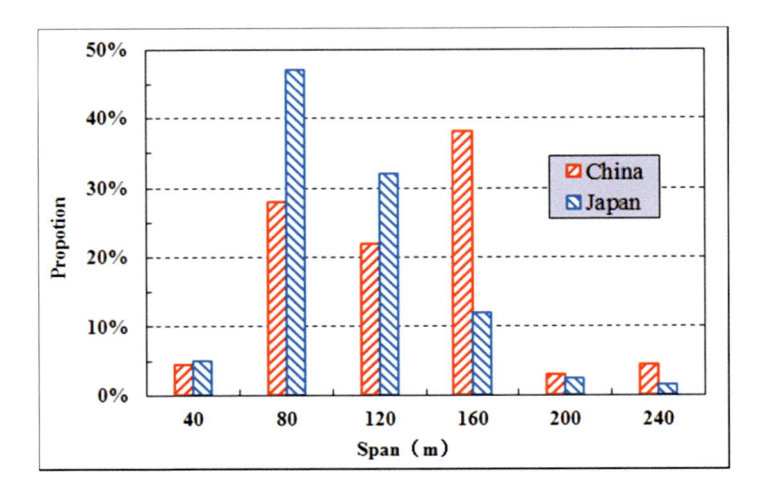

Figure 2.14 RW cantilever construction. *RW*, ripple web.

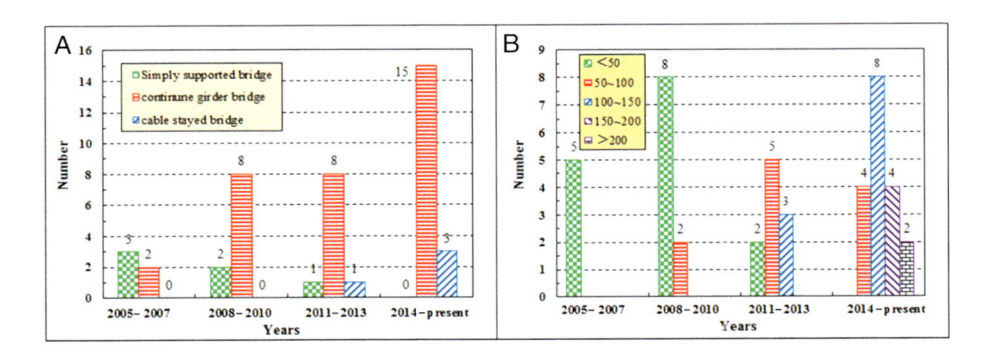

Figure 2.15 Developmental trends of bridge with CSWs: (A) trends of bridge type; (B) trends of bridge span. *CSWs*, corrugated steel webs.

and small span bridges (span between 50 ~ 100m). Conversely, large-span bridges with CSW, with spans over 150 ~ 200m, have been built in China. The largest span continuous girder (Qianshanhe River Bridge) and extradosed bridge (Chaoyanggou Bridge) with CSWs all over the world are built in China.

Fig. 2.15 shows the developmental trends of these new-style composite bridges with CSWs in the last 15 years in China. Initially, most bridges had spans less than 50 m and they were almost simply supported bridges. Then, with the accumulation of engineering experience and the development of design technology, composite bridges with CSWs have been developed to reach large spans and to cover complex bridge systems, such as cable-stayed bridges.

References

[1] R.J. Jiang, F.T.K. Au, Y.F. Xiao, Prestressed concrete girder bridges with corrugated steel webs: review, J. Struct. Eng. 141 (2) (2014) 04014108.

[2] A.H. Yuan, J.B. Chen, S. Wan, B.C. Lu, On a design of pre-stressed concrete composite continuous box-girder footbridge with corrugated steel webs, J. Univ. Sci. Technol. Suzhou (Engineering and Technology) 17 (3) (2004) 55–58 (in Chinese).

[3] S.F. Wang, The main bridge of Qianshan Bridge cantilever steel webs beam construction, Construct. Des. Project 17 (2016) 143–145 in Chinese.

[4] S.S. Chen, Z.B. Zhong, S.R. Gui, X.Z. Liu, H.Q. Zhong, Structural design of multi-pylon cable-stayed bridge with corrugated steel webs for Chaoyang bridge in Nanchang, World Bridge 6 (2014) 1–6 in Chinese.

[5] M. Zhou, Z. Liu, J. Zhang, L. An, Z. He, Equivalent computational models and deflection calculation methods of box girders with corrugated steel webs, Eng. Struct. 127 (2016) 615–634.

[6] M. Zhou, Z. Liu, J. Zhang, L. An, Deformation analysis of a non-prismatic beam with corrugated steel webs in the elastic stage, Thin-Walled Struct. 109 (2016) 260–270.

[7] M. Zhou, J. Zhang, J. Zhong, L. An, Shear stress calculation and distribution in variable cross sections of box girders with corrugated steel webs, J. Struct. Eng. 142 (6) (2016) 04016022.

[8] M. Zhou, Z. Liu, J. Zhang, H. Shirato, Stress analysis of linear elastic nonprismatic concrete-encased beams with corrugated steel webs, J. Bridge Eng. 22 (6) (2017) 04017012.

[9] J. Nie, L. Zhu, M. Tao, Shear strength of trapezoidal corrugated steel webs, China Civil Eng. J. 85 (6) (2013) 105–115 in Chinese.

[10] J. Nie, Y. Zhu, M. Tao, C. Guo, Y. Li, Optimized prestressed continuous composite girder bridges with corrugated steel webs, J. Bridge Eng. 22 (2) (2016) 04016121.

[11] J. He, Y. Liu, A. Chen, T. Yoda, Mechanical behavior and analysis of composite bridges with corrugated steel webs: state-of-the-art, Int. J. Steel Struct. 12 (3) (2012) 321–338.

[12] D. Jin, X. Zhou, X. Kong, D. Cheng, Z. Qian, Experimental research on pre-stressed concrete composite box girders with corrugated steel webs, J. Changan Univ. 29 (5) (2009) 64–70 in Chinese.

[13] L. Li, Z. Liu, F. Wang, Theoretical and experimental research on the flexural behavior of external prestressed composite beam with corrugated webs, Eng. Mech. 26 (7) (2009) 89–96 in Chinese.

[14] Y. Khalid., C. Chan, B. Sahari, A. Hamoud, Bending behaviour of corrugated web beams, J. Mater. Process. Technol. 150 (2004) 242–254.

[15] J. He, Y. Liu, A. Chen, D. Wang, T. Yoda, Bending behavior of concrete-encased composite I-girder with corrugated steel web, Thin-Walled Struct. 74 (2014) 70–84.

[16] H. Abbas, R. Driver, R. Sause, Shear behavior of corrugated web bridge girders, J. Struct. Eng. 132 (2) (2006) 195–203.

[17] Y.Q. Liu, Hybrid Bridge, China Communication Press, Beijing, 2005.

[18] K.B. Jiang, Y. Ding, Y.W. Liu, F. Zheng, Calculation for torsion strength of prestressed concrete beams based on fixed-angle softened truss model, Appl. Mech. Mater. 52-54 (2011) 1032–1038.

[19] J. Yi, H. Gil, K. Youm, H. Lee, Interactive shear buckling behavior of trapezoidally corrugated steel webs, Eng. Struct. 30 (6) (2008) 1659–1666.

[20] J. Di, X. Zhou, P. Qiao, X. Kong, K. Yu, J. Hou, Shear buckling property of corrugated steel webs, J. Traffic Transport. Eng. 3 (9) (2009) 11–18 in Chinese.

[21] E.C. Oguejiofor, M.U. Hosain, A parametric head study of perfobond rib shear connectors, Can. J. Civ. Eng. 21 (4) (1994) 614–625.

[22] E.C. Oguejiofor, M.U. Hosain, Numerical analysis of push-out specimens with perfobond rib connectors, Comput. Struct. 62 (4) (1997) 617–624.

[23] J.H. Ahn, S.H. Kim, Y.J. Jeong, Fatigue experiment of stud welded on steel plate for a new bridge deck system, Steel Composite Struct. 7 (5) (2007) 391–404.

[24] J.H. Ahn, C.G. Lee, J.H. Won, S.H. Kim, Shear resistance of the perfobond-rib shear connector depending on concrete strength and rib arrangement, J. Constr. Steel Res. 66 (10) (2010) 1295–1307.

[25] Z.H. Liang, S.F. Yuan, C.X. Hu, Construction techniques for corrugated steel web and pc composite girder of Juancheng Huanghe River highway bridge, Bridge Construct. 6 (2010) 73–76 in Chinese.

[26] S.M. Zheng, W.K. Xiang, B.W. Yang, S. Wan, Study of positioning of spatial location of corrugated steel webs in construction, Bridge Construct. 42 (4) (2012) 113–117 in Chinese.

[27] G.J. Feng, H. Chen, J. Li, D.H. Chen, Analysis of load bearing performance of corrugated steel web extradosed bridge during construction process, World Bridge 44 (3) (2016) 48–52 in Chinese.

[28] Y.J. Du, B.W. Yang, S. Wan, Installation technology of webs in the construction of pc box-girder bridge with corrugated steel webs, Appl. Mech. Mater. 204-208 (2012) 2209–2213.

[29] K.B. Jiang, Y. Ding, Y.W. Liu, P. Wang, F. Zheng, Sensitivity analysis for design parameters of cantilever casting pc box-girder with corrugated steel webs, Adv. Mater. Res. 430-432 (2012) 1546–1550.

[30] D. Wang, H.S. Huang, Z. Cao, Y. Liu, Analysis of mechanical property of composite box-girder with corrugated steel webs using new asynchronous construction technology, J. Highway Transport. Res. Develop. 33 (8) (2016) 58–64 in Chinese.

[31] K.S. Kim, K.H. Jung, C.W. Sim, J.G. Han, Design and construction of hybrid bridge with corrugated steel web by incremental launching method, Civil Environ. Eng. 5 (2005) 411–414 in Korean.

[32] K.H. Jung, K.S. Kim, C.W. Sim, J.H.J. Kim, Verification of incremental launching construction safety for the Ilsun Bridge: The world's longest and widest prestressed concrete box girder with corrugated steel web section, J. Bridge Eng. 16 (3) (2011) 453–460.

Real boundary condition between flange and web

3.1 Scope

Although the concept of the real boundary condition at the juncture of the web and flanges of I-plate girders with *flat webs* to be somewhere between simple and fixed has been recognized from early days, the boundary condition has been conservatively assumed simple, mainly due to the lack of means to evaluate it in a rational manner [1]. Therefore, a series of numerical analyses was performed by Lee et al. [1] in 1996, based on a successfully verified three-dimensional finite element (FE) modeling, to investigate the effects of the geometric parameters on the boundary conditions at the web-to-flange juncture. It was concluded that the realistic support condition at the juncture is closer to a fixed support for the geometric parameters that are associated with practical designs; $t_f/t_w > 2.0$, where t_f and t_w are the flange and web thicknesses, respectively.

On the other hand, to the authors' best knowledge, *the real behavior at the juncture between the corrugated web (CW) and the flanges of bridge girders* has never been addressed in design standards. Accordingly, this chapter aims to shed light on the real boundary conditions of corrugated web girders (CWGs). To do so, elastic bifurcation buckling analyses are performed, using ABAQUS *FE* package [2], on CWs with simple (S) and fixed (F) boundary conditions at their juncture with the flanges. The considered CWs had average dimensions of the available bridge girders with CWs. Then, the behavior of each type (simple (S) and fixed (F) boundary conditions) is checked by comparing them with each other. After that, the validity of the available interactive shear buckling strength formulas is examined with the critical shear buckling strengths of the CWs with both boundary conditions. Finally, a comparison between the girder segments of CWs and flanges with the CWs with simple (S) and fixed (F) boundary conditions is carried out, in order to define the limit of t_f/t_w ratio between the simple (S) and fixed (F) juncture behavior.

3.2 Background

As shown in the previous chapter, bridges with corrugated steel webs have been extensively constructed in different countries. A trapezoidally corrugated steel plate, as shown in Fig. 3.1, is composed of a series of longitudinal and inclined sub-panels. The corrugations of I-section girders provide shear stability to the webs through increasing their out-of-plane stiffness, eliminating the need for transverse stiffeners which have a principal influence on the shear strength of conventionally stiffened flat-webbed I-section plate girders. Accordingly, they provide relatively high strength-to-weight

Behavior and Design of Trapezoidally Corrugated Web Girders for Bridge Construction: Recent Advances.
DOI: https://doi.org/10.1016/B978-0-323-88437-2.00005-8

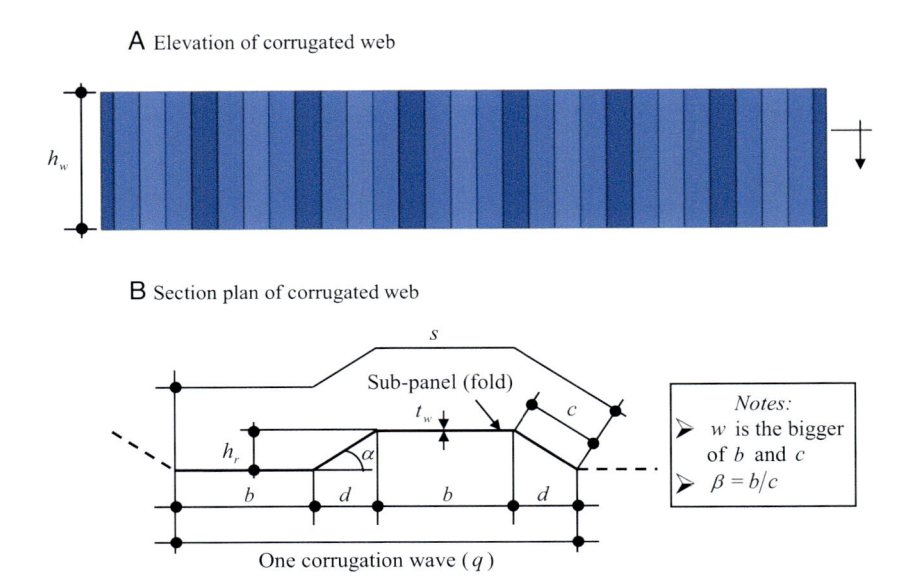

Figure 3.1 Corrugation configuration and geometric notation.

ratios. The geometric properties of a number of bridges with trapezoidally corrugated steel web plates [3–5] are provided in Table 3.1.

For a steel CWG, it is assumed that the web carries merely shear forces due to the accordion effect [6–7]. The web in such girder was found to carry insignificant longitudinal stresses from the primary flexure. Consequently, the bending moment is nearly carried by the flanges. Thus, it is worth mentioning that the shear strength can be determined without consideration of moment-shear interaction. Accordingly, the web shear stress (τ) can be assumed constant and quantified in terms of the average shear stress, as follows:

$$\tau = \frac{V}{t_w h_w} \tag{3.1}$$

where V is the vertical shear force in the girder, h_w is the web depth, and t_w is the web thickness. Shear buckling mechanism in steel CWs is classified as *local, global,* or *interactive* buckling. The first buckling mode is controlled by deformations within a single sub-panel (fold) of the web. The second buckling mode involves multiple folds and the buckled shape extends diagonally over the depth of the web. However, experimentally observed buckling *often* appears to have characteristics of both local and global buckling modes. This was classified as an interactive buckling mode. Historically, research on the shear buckling behavior of corrugated plates has been initiated in 1969 by Easley and McFarland [8]. Initially, the shear strength and behavior of trapezoidal corrugated steel web plates has extensively been investigated experimentally. For instance, Lindner and Aschinger [9], Elgaaly et al. [10], El-Metwally [11], Abbas [12], Sause et al. [13], Driver et al. [14], and Moon et al. [4] presented

Table 3.1 Profiles of corrugated steel webs of constructed bridges [3–5] and their geometric properties.

Bridge name	b [mm]	d [mm]	c [mm]	h_r [mm]	t_w [mm]	h_w [mm]	s [mm]	q [mm]	α [°]	β	w/h_w [a]	h_r/t_w
Shinkai	250	200	250	150	9	1183	1000	900	36.9	1.00	0.21	16.7
Matsnoki	300	260	300	150	10	2210	1200	1120	30.0	1.00	0.14	15.0
Hondani	330	270	336	200	9	3315	1332	1200	36.5	0.98	0.10	22.2
Cognac	353	319	353	150	8	1771	1412	1344	25.2	1.00	0.20	18.8
Maupré	284	241	284	150	8	2650	1136	1050	31.9	1.00	0.11	18.8
Dole	430	370	430	220	10	2546	1720	1600	30.7	1.00	0.17	22.0
Ilsun	330	330	386	200	18	2292	1432	1320	31.2	0.85	0.17	11.1
Ave	325	284	334	174	10	2281	1319	1219	32	0.98	0.16	17.8

[a] w is the bigger of b and c.

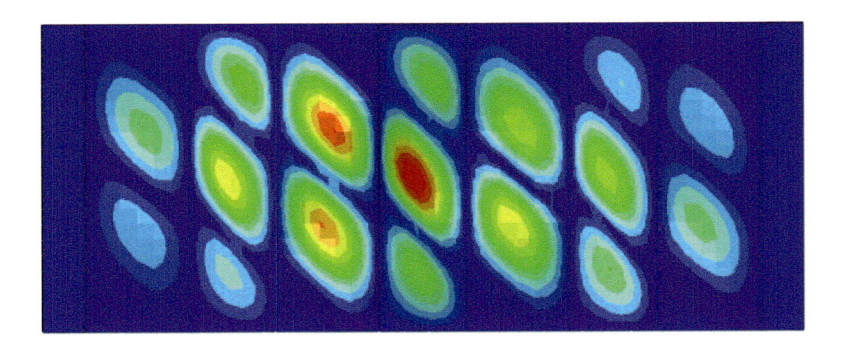

Figure 3.2 Local shear buckling mode (from the current modeling).

test results for the shear strength of steel trapezoidal CWs. These experimental studies were followed by analytical investigations that *well simulate* the actual behavior under shear of the trapezoidally corrugated steel web plates. A sample of these analytical investigations and discussions were presented by Yi et al. [3], Moon et al. [4], and more recently by Sause and Braxtan [15].

3.3   Elastic shear buckling behavior

In the following, a brief discussion about the three buckling modes of the CWs (local, global and interactive modes) is provided.

3.3.1   Local shear buckling

Local buckling, as can be seen in Fig. 3.2, is controlled by the slenderness of the individual folds of the web. It occurs when a flat sub-plate or fold between vertical edges has a large width-to-thickness ratio. In this mode, the CW acts as a series of flat plate sub-panels that mutually support each other along their vertical edges and are supported by the flanges at their shorter horizontal edges.

The elastic local shear buckling stress of the CWs ($\tau_{cr,L}$) can be determined by the classical *plate buckling theory,* which is based on results of many investigators including Timoshenko and Gere [16], Bulson [17], and Galambos [18], which is expressed as:

$$\tau_{cr,L} = k_L \frac{\pi^2 E}{12(1 - \upsilon^2)} \left(\frac{t_w}{w}\right)^2 \tag{3.2}$$

where E is Young's modulus of elasticity; υ is the Poisson's ratio; w is the maximum fold width (maximum of flat panel width b, and inclined panel width c); t_w is the web

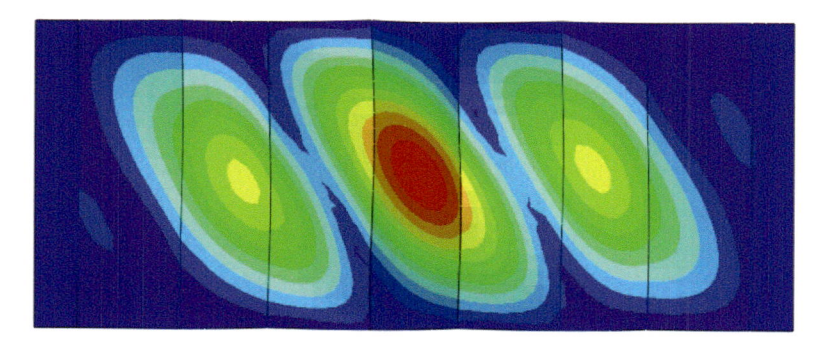

Figure 3.3 Global shear buckling mode (from the current modeling).

thickness; and k_L is the local shear buckling coefficient which is a function of the aspect ratio of the sub-panel (w/h_w), as follows:

$$k_L = 5.34 + 4\left(\frac{w}{h_w}\right)^2 \tag{3.3}$$

For CWs with constant height (h_w) and thickness (t_w), the parameter β, which is the ratio of the length of the longitudinal sub-panel (b) to the length of the inclined sub-panel (c), represents the tendency for the local buckling of a wider fold to be restrained by adjacent folds. When $\beta = 1.0$, adjacent sub-panels are of the same width and are equally critical for local shear buckling; when $\beta > 1.0$, the longitudinal sub-panels are wider and more critical for local shear buckling and are restrained by the inclined folds; and when $\beta < 1.0$, the inclined sub-panels are wider and more critical for local shear buckling and are restrained by the longitudinal folds. Hence, according to Sause and Braxtan [15], the local buckling shear strength formulas apply only when $\beta = 1.0$, and for other cases, the actual local buckling shear strength should be greater than that estimated by the shear strength formula of Eq. (3.2).

3.3.2 Global shear buckling

Fig. 3.3 represents the global buckling mode. When global buckling controls the failure mode, the buckling stress can be calculated for the whole CW panel using the *orthotropic-plate buckling theory*; refer to Galambos [18]. Historically, the calculation of the global elastic buckling stress ($\tau_{cr,G}$) for the CWs was initiated by Easley [19] in 1975, as follows:

$$\tau_{cr,G} = k_G \frac{D_x^{0.25} D_y^{0.75}}{t_w h_w^2} \tag{3.4}$$

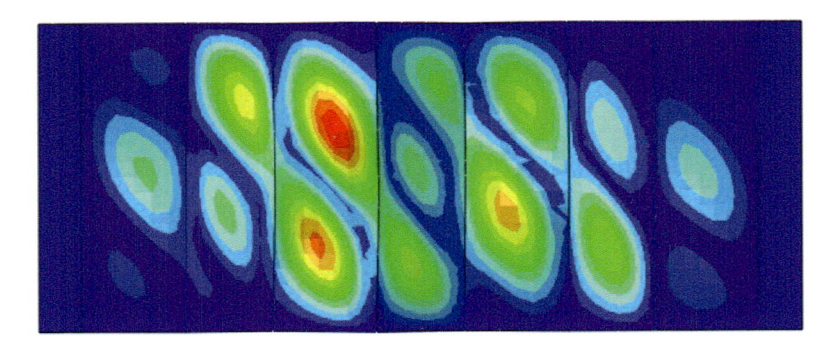

Figure 3.4 Interactive shear buckling mode (from the current modeling).

where the transverse bending stiffness per unit length of the CW (D_x), the longitudinal bending stiffness per unit length of the CW (D_y), and I_y are defined as:

$$D_x = \frac{q}{s} \cdot \frac{E t_w^3}{12} \tag{3.5}$$

$$D_y = \frac{E I_y}{q} \tag{3.6}$$

$$I_y = 2 b t_w \left(\frac{h_r}{2}\right)^2 + \frac{t_w h_r^3}{6 \sin \alpha} \tag{3.7}$$

k_G is the global buckling coefficient. Elgaaly et al. [10] assume that the web is relatively long compared to h_w and suggest that k_G is to be taken as 31.6 (assuming the web is simply supported by the flanges) or 59.2 (assuming that the flanges provide the web with fixed supports). Easely [19], however, suggests that k_G varies between 36 and 68.4.

However, several derivations from Eqs. (3.4–3.7) for the calculation of the global shear buckling stress were made by other researchers. For example, Abbas [12] expressed the global shear buckling stress directly in terms of the geometric parameters of the trapezoidal corrugations. Another expression for $\tau_{cr,G,}$ based also on Eqs. (3.4–3.7), was provided by Yi et al. [3].

3.3.3 Interactive shear buckling

Test specimens and FE models with characteristics of both local and global buckling modes were observed in the literature [3–8]. These shear buckling modes (Fig. 3.4), termed interactive buckling, are explained as a result of the interaction between local and global buckling. Interaction formulas, originally proposed by Lindner and Aschinger [9], are generalized as follows:

$$\frac{1}{(\tau_{cr,I})^n} = \frac{1}{(\tau_{cr,L})^n} + \frac{1}{(\tau_{cr,G})^n} \tag{3.8}$$

Table 3.2 Interactive shear buckling strength formulas.

Paper	Year	Interactive shear buckling strength formulas
Bergfelt and Leiva [20] Yi et al. [3]	1984 2008	$\dfrac{1}{(\tau_{cr,I})} = \dfrac{1}{(\tau_{cr,L})} + \dfrac{1}{(\tau_{cr,G})}$
El-Metwally [11]	1998	$\dfrac{1}{(\tau_{cr,I})^2} = \dfrac{1}{(\tau_{cr,L})^2} + \dfrac{1}{(\tau_{cr,G})^2} + \dfrac{1}{(\tau_y)^2}$
Abbas et al [21]	2002	$\dfrac{1}{(\tau_{cr,I})^2} = \dfrac{1}{(\tau_{cr,L})^2} + \dfrac{1}{(\tau_{cr,G})^2}$
Shiratoni [22]	2003	$\dfrac{1}{(\tau_{cr,I})^4} = \dfrac{1}{(\tau_{cr,L})^4} + \dfrac{1}{(\tau_{cr,G})^4}$
Sayed-Ahmed [23]	2005	$\dfrac{1}{(\tau_{cr,I})^3} = \dfrac{1}{(\tau_{cr,L})^3} + \dfrac{1}{(\tau_{cr,G})^3} + \dfrac{1}{(\tau_y)^3}$

which is solved as:

$$\tau_{cr,I} = \frac{\tau_{cr,L} \cdot \tau_{cr,G}}{\left((\tau_{cr,L})^n + (\tau_{cr,G})^n\right)^{\frac{1}{n}}} \tag{3.9}$$

Several researches were conducted to find the best exponent (n) of Eq. (3.9). Table 3.2 provides the previously proposed interactive shear buckling strength formulas [3,11,20–23]. It should be noted that the interaction among local buckling, global buckling, and yield strength has been used in some cases.

3.4 Current finite element models

The real behavior at the juncture of the web and flanges is examined herein for the case of trapezoidally steel CW plates. However, in order to calculate the elastic buckling strength, the boundary conditions of the web panel that is stiffened by transverse stiffeners should be determined. It is commonly assumed that transverse stiffeners are sufficiently stiff to form nodal lines of sinusoidal waves of buckled modes on the web, and they are always designed to satisfy this condition [1]. This implies that the stiffeners will not deform during buckling. Conversely, the web panel is elastically restrained at the juncture between the web and flanges.

One hundred and sixty-eight FE models were generated using ABAQUS FE program [2] on trapezoidally corrugated steel web plate segments covering five parameters. It is expected, based upon the authors' experience, that the most relevant parameters influencing the shear response of the trapezoidally corrugated steel web plate segments are the following:

1. Corrugation depth-to-web thickness ratio (h_r/t_w); (9.72–29.17).
2. Web flat panel width-to-depth ratio (b/h_w); (0.135–0.325).
3. Aspect ratio of the web panel (a/h_w); (1.0–2.4).
4. Web plate slenderness (h_w/t_w); (56–400).
5. Flange thickness-to-web thickness ratio (t_f/t_w).

Each segment consists of two corrugation waves ($a = 4b + 4d$). The dimensions of the corrugations were assumed here by using the average value of flat plate width ($b = 325$ mm) for the available bridges, as can be seen in Table 3.1. The CW was then assumed to have a (b/c) ratio of 1.0, as noted by Sause and Braxton [15] that this aspect ratio is the most critical. The corrugation depth (h_r) of 175 mm is also the average value of the depth used in the available bridges. The longitudinal projection of the inclined panel (d) was consequently found to be 274 mm. The web thickness (t_w) was then assumed to accord with the results of Easley [19] by meeting the desired geometric stiffness of $h_r/t_w > 10$.

Based on a preliminary analysis on the effect of flange width on the CW shear critical stress as can be seen later in this chapter, one flange width of 400 mm was used in the current study for the case of girder segments. The web depth of the girders was varied from 1000 mm to 2400 mm for the entire FE program insuring that the aspect ratio of the web panel (a/h_w) is greater than and including 1.0. Considering that $a/h_w \geq 1.0$ is consistent with the conditions of CWGs used in practice, where the distance between vertical stiffeners is much greater than the web depth (h_w). Additionally, this condition is consistent with the theory behind the shear strength formulas, which assumes a long web relative to h_w with simply supported boundary conditions [10,15].

The CWs of the girder segments were stiffened transversely at their ends with sufficiently stiff stiffeners to provide nodal lines of sinusoidal waves of buckled modes on the web [1]. Each girder segment was subjected to a concentrated load, as can be seen in Fig. 3.5, which provides a pure shear loading.

On the other hand, it is well known that the effect of cold-forming process of the corrugations increases the yield strength of the base material in a small region around the corner of the corrugation [24]. However, the corner effects due to cold-forming were found to have insignificant effect on the ultimate strength of the girders [24] and consequently on the critical shear buckling. Accordingly, this issue was not included in the present modeling.

The trapezoidally corrugated steel web plate segments are labeled such that the juncture between the web and flanges: S, F, RF representing the simple, fixed, and real flange condition (girder segment), respectively, could be identified from the label. The details of the groups are given in Table 3.3.

3.5 Verification of finite element models

The three-dimensional FE simulation permits a closer examination of the real behavior of the juncture of the web and flanges. Therefore, an elastic bifurcation buckling analysis, using the general-purpose program ABAQUS [2], was made for isolated trapezoidally corrugated steel web plate segments between two adjacent transversal stiffeners for the extreme boundary conditions; simple (S) or fixed (F). To allow the comparison, girder segments of trapezoidally corrugated steel webs with flat flange plates (RF) were in addition simulated. Following Yi et al. [3], S8R5 reduced integration thin shell elements were used in the current bifurcation buckling analysis. To insure that the proposed FE simulation adequately represents the actual elastic buckling strength,

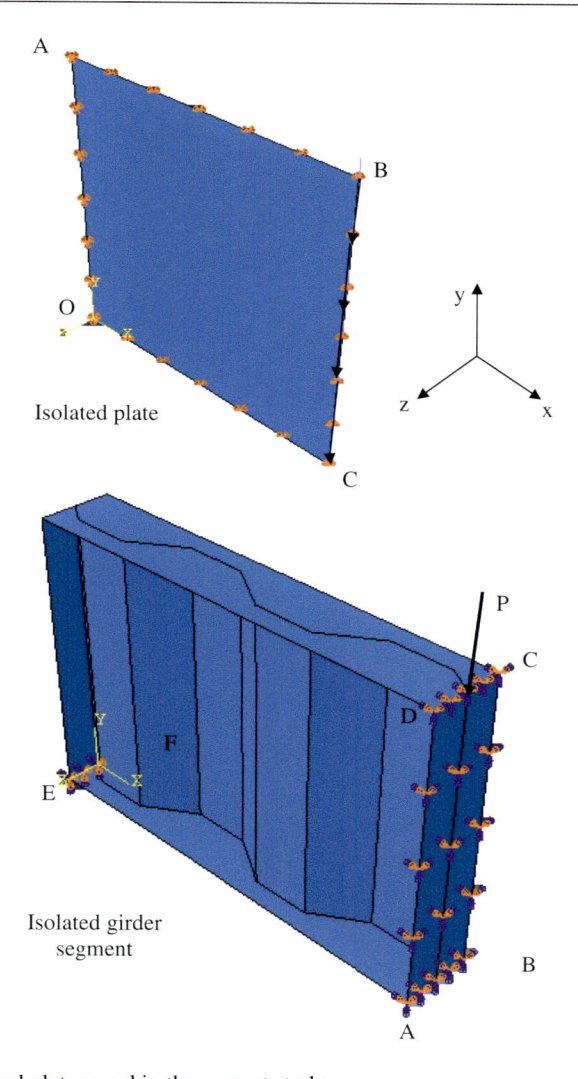

Figure 3.5 Isolated plates used in the current study.

a convergence test has been carried out for a *square isolated flat plate* (1500 mm × 1500 mm × 6 mm) subjected to pure shear. The four edges of the isolated plate were considered simple, as can be seen in Table 3.4 and Fig. 3.5. The (R) represents a restrained boundary condition, while (F), in Table 3.4, stands for a free boundary condition. Young's modulus of 210 GPa and a Poisson's ratio of 0.3 were used during the analysis. The buckling shapes are plotted in Fig. 3.6. The results of the convergence test can be viewed in Fig. 3.7, which represents the relationship between the ratio of the *FE* shear-buckling stress to its elastic critical value (τ_{FE}/τ_{cr}) versus the number of elements per horizontal width. The elastic shear-buckling stress (τ_{cr}) was calculated

Table 3.3 Current finite element program.

Cross-section of flange[a]	h_w [m]	t_w [mm]	t_f [mm]	$\dfrac{a}{h_w}$	$\dfrac{h_w}{t_w}$	$\dfrac{t_f}{t_w}$	$\dfrac{b}{h_w}$	$\dfrac{h_r}{t_w}$
S	1.0–2.4	6–18	–	1.0–2.4	56–400	–	0.135–0.325	9.72–29.17
F	1.0–2.4	6–18	–	1.0–2.4	56–400	–	0.135–0.325	9.72–29.17
RF	1.0–2.4	6–18	18–90	1.0–2.4	56–400	1–5	0.135–0.325	9.72–29.17

[a] S, simple boundary condition without flanges; F, fixed boundary condition without flanges; RF, real flange condition.

Table 3.4 Boundary conditions used for the isolated web plates and girder segments.

Deformation	symbols	Simple isolated plate				Fixed isolated plate				Isolated girder segment	
		OA	AB	BC	CO	OA	AB	BC	CO	ABCD	Edge EF
Translation	δ_x	R	R	R	R	R	R	R	R	R	F
	δ_y	R	F	F	F	R	F	F	F	F	R
	δ_z	R	R	R	R	R	R	R	R	R	R
Rotation	θ_x	F	F	F	F	F	R	F	R	R	R
	θ_y	F	F	F	F	F	R	F	R	R	R
	θ_z	F	F	F	F	F	R	F	R	R	F

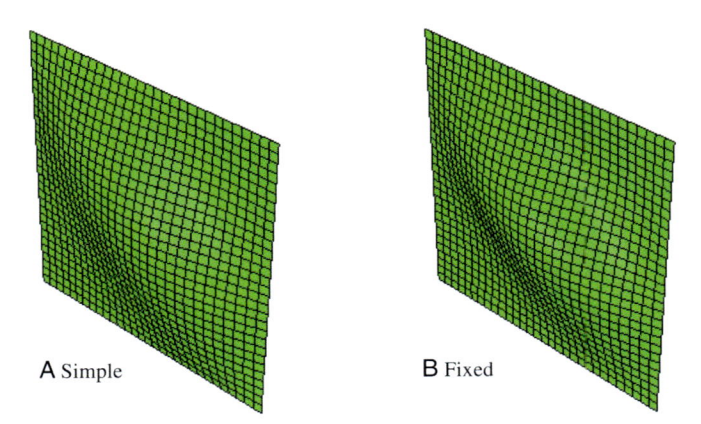

A Simple   B Fixed

Figure 3.6 Buckling shape for (A) simple and (B) fixed isolated plates.

using the classical *plate buckling theory* [16] which is expressed as:

$$\tau_{cr} = \frac{k\pi^2 E}{12(1 - v^2)}\left(\frac{t_w}{h_w}\right)^2 \tag{3.10}$$

where E is Young's modulus of elasticity of 210 GPa; v is the Poisson's ratio of 0.3; t_w is the web thickness (6 mm); h_w is the web depth of 1500 mm; and k is the shear-buckling coefficient equaling to 9.34 for a square simply supported plate.

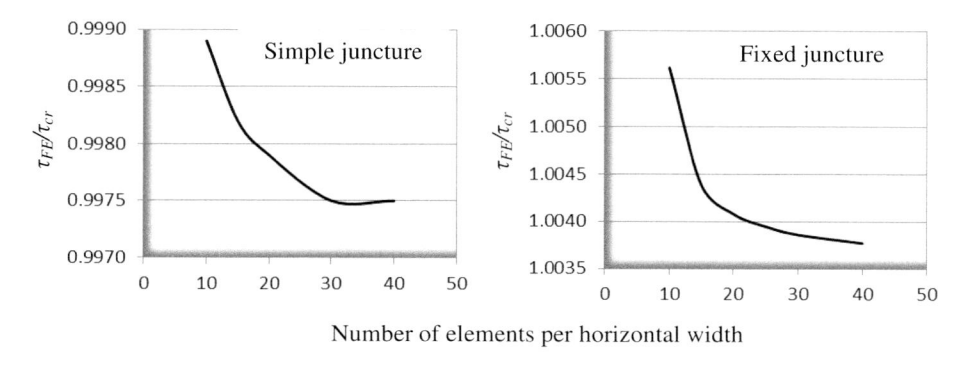

Number of elements per horizontal width

Figure 3.7 Results of convergence test.

For the case of fixed boundary condition at the juncture of the web and flanges, a convergence test was in addition conducted based upon the boundary conditions given in Table 3.4. The results of convergence test can as well be viewed in Fig. 3.7, which represents the relationship between the ratio of the *FE* shear-buckling stress to its elastic critical value (τ_{FE}/τ_{cr}) versus the number of elements per horizontal width. The elastic shear-buckling stress (τ_{cr}) for the current case was calculated using the classical *plate buckling theory* [16], as previously given in Eq. (3.10) by using the shear buckling coefficients (k_{sf}) according to Bulson [17], as follows:

$$k_{sf} = 8.98 + \frac{5.61}{(a/h_w)^2} - \frac{1.99}{(a/h_w)^3} \; if \; a/h_w \geq 1.0 \tag{3.11}$$

On the other hand, the boundary conditions of the current models for simulating the girder segments of trapezoidally corrugated steel webs elastic buckling behavior were then simulated following Eldib [5], as can be seen in Table 3.4 and Fig. 3.5. It should be mentioned that this model was verified by Moon et al. [4] using the experimental test results of other researchers.

From the previous convergence tests, it can be seen that the results of the *FE* modeling are close to the results of the *plate buckling theory* [16]. Hence, it can be concluded that the previous *FE* modeling is *suitable* to simulate the shear behavior of flat webs. Consequently, it can be ensured that using this technique for the case of CWs leads to trusted results, typically as made by Yi et al. [3].

3.6 General results

The main objective of the current section is to investigate the extreme boundary conditions of the juncture between the CW and flanges of bridge girders; simple (S) and fixed (F). Accordingly, this section provides the results of the current trapezoidally corrugated steel web plates and their discussions. The critical shear stress ($\tau_{cr,FE}$) for each model was determined from the results of the *FE* modeling, as can be seen in Table 3.5 for the case of simple (S) and fixed (F) boundary conditions.

Table 3.5 Results of the finite element program for models of simple (S) and fixed (F) juncture between the CW and flanges.

No.	h_w [mm]	t_w [mm]	h_w/t_w	a/h_w	b/h_w	h_r/t_w	Buckling mode[a]		$\tau_{cr,FE}$ [N/mm^2]		[10]/[11]
							S	F	S	F	
[1]	[2]	[3]	[4]	[5]	[6]	[7]	[8]	[9]	[10]	[11]	[12]
1	1000	6	167	2.40	0.325	29.17	L	L	366	377	0.97
2	1000	8	125	2.40	0.325	21.88	I	I	612	636	0.96
3	1000	10	100	2.40	0.325	17.50	I	I	839	932	0.95
4	1000	12	83	2.40	0.325	14.58	I	I	1184	1255	0.94
5	1000	14	71	2.40	0.325	12.50	I	I	1489	1579	0.94
6	1000	16	63	2.40	0.325	10.94	I	I	1782	1904	0.94
7	1000	18	56	2.40	0.325	9.72	I	I	2085	2235	0.93
8	1200	6	200	2.00	0.271	29.17	I	I	355	362	0.98
9	1200	8	150	2.00	0.271	21.88	I	I	589	605	0.97
10	1200	10	120	2.00	0.271	17.50	I	I	846	880	0.96
11	1200	12	100	2.00	0.271	14.58	I	I	1116	1173	0.95
12	1200	14	86	2.00	0.271	12.50	I	I	1421	1460	0.97
13	1200	16	75	2.00	0.271	10.94	I	I	1692	1755	0.96
14	1200	18	67	2.00	0.271	9.72	G	I	1954	2059	0.95
15	1400	6	233	1.71	0.232	29.17	I	I	351	354	0.99
16	1400	8	175	1.71	0.232	21.88	I	I	578	590	0.98
17	1400	10	140	1.71	0.232	17.50	I	I	818	853	0.96
18	1400	12	117	1.71	0.232	14.58	G	I	1070	1119	0.96
19	1400	14	100	1.71	0.232	12.50	G	I	1329	1391	0.96
20	1400	16	88	1.71	0.232	10.94	G	I	1594	1673	0.95
21	1400	18	78	1.71	0.232	9.72	G	I	1864	1962	0.95
22	1600	6	267	1.50	0.203	29.17	L	L	350	354	0.99
23	1600	8	200	1.50	0.203	21.88	I	I	571	584	0.98
24	1600	10	160	1.50	0.203	17.50	G	I	798	835	0.96
25	1600	12	133	1.50	0.203	14.58	G	I	1039	1086	0.96
26	1600	14	114	1.50	0.203	12.50	G	I	1291	1351	0.96
27	1600	16	100	1.50	0.203	10.94	G	I	1548	1624	0.95
28	1600	18	89	1.50	0.203	9.72	G	I	1809	1935	0.93
29	1800	6	300	1.33	0.181	29.17	L	L	347	351	0.99
30	1800	8	225	1.33	0.181	21.88	G	I	562	585	0.96
31	1800	10	180	1.33	0.181	17.50	G	I	783	820	0.95
32	1800	12	150	1.33	0.181	14.58	G	I	1020	1067	0.96
33	1800	14	129	1.33	0.181	12.50	G	G	1264	1324	0.95
34	1800	16	113	1.33	0.181	10.94	G	G	1510	1572	0.96
35	1800	18	100	1.33	0.181	9.72	G	G	1758	1822	0.96
36	2000	6	333	1.20	0.163	29.17	I	I	343	350	0.98
37	2000	8	250	1.20	0.163	21.88	G	G	553	579	0.96
38	2000	10	200	1.20	0.163	17.50	G	G	774	810	0.96
39	2000	12	167	1.20	0.163	14.58	G	G	1004	1047	0.96
40	2000	14	143	1.20	0.163	12.50	G	G	1236	1280	0.97
41	2000	16	125	1.20	0.163	10.94	G	G	1469	1515	0.97

(*continued on next page*)

Table 3.5 Results of the finite element program for models of simple (S) and fixed (F) juncture between the CW and flanges—cont'd

No.	h_w [mm]	t_w [mm]	h_w/t_w	a/h_w	b/h_w	h_r/t_w	Buckling mode[a]		$\tau_{cr,FE}$ [N/mm²]		[10]/[11]
							S	F	S	F	
42	2000	18	111	1.20	0.163	9.72	G	G	1705	1754	0.97
43	2200	6	367	1.09	0.148	29.17	I	I	341	345	0.99
44	2200	8	275	1.09	0.148	21.88	G	G	549	572	0.96
45	2200	10	220	1.09	0.148	17.50	G	G	765	796	0.96
46	2200	12	183	1.09	0.148	14.58	G	G	983	1017	0.97
47	2200	14	157	1.09	0.148	12.50	G	G	1202	1237	0.97
48	2200	16	138	1.09	0.148	10.94	G	G	1425	1463	0.97
49	2200	18	122	1.09	0.148	9.72	G	G	1655	1694	0.98
50	2400	6	400	1.00	0.135	29.17	I	I	344	344	1.00
51	2400	8	300	1.00	0.135	21.88	G	G	547	567	0.96
52	2400	10	240	1.00	0.135	17.50	G	G	753	778	0.97
53	2400	12	200	1.00	0.135	14.58	G	G	958	985	0.97
54	2400	14	171	1.00	0.135	12.50	G	G	1168	1197	0.98
55	2400	16	150	1.00	0.135	10.94	G	G	1385	1416	0.98
56	2400	18	133	1.00	0.135	9.72	G	G	1610	1644	0.98
									Ave		**0.96**
									COV		**0.014**

[a] Buckling modes: Local (L), Global (G), and Interactive (I).

It can be noticed that the critical shear stress ($\tau_{cr,FE}$) of the CWs of simple juncture with the flanges varies between 0.93 to 1.0 from that of the corresponding fixed boundary condition with an average value of 0.96.

3.7 Comparison between finite element critical stresses and available formulas

The critical shear stress ($\tau_{cr,FE}$) of the CWs of both simple and fixed junctures are compared in this section with first-order interactive buckling strength ($\tau_{cr,I,1}$), proposed by Yi. et al [3], and presented in Eq. (3.12). The results of comparison are provided in Table 3.6. Two values for the global shear buckling coefficient (k_G) of 36 [19] and 31.6 [10] were used for the case of simple juncture. For the fixed juncture case, also two values were checked; 68.4 [19] and 59.2 [10]. From the comparative results, it can be noticed that the first-order interactive buckling strength ($\tau_{cr,I,1}$) is *suitable* for the case of CWs with simple juncture using both values of k_G. On the other hand, the same first-order interactive buckling strength ($\tau_{cr,I}$) seems to be *unconservative* for the case of fixed juncture CWs. It should be mentioned that the interactive buckling strengths ($\tau_{cr,I}$) of higher values of (n) were also calculated according to [11,21–23] and the relative strengths were found to be *more unconservation* compared to Yi et al. [3]. For example, the average value of the critical shear stress to the second-order interactive

Table 3.6 Comparison between the critical shear stress ($\tau_{cr,FE}$) of the CWs of simple and fixed juncture with first-order interactive buckling strength [3].

			S		F		F	
			$\tau_{cr,FE}/\tau_{cr,I,1}$		$\tau_{cr,FE}/\tau_{cr,I,1}$		$\tau_{cr,FE}/\tau_{cr,I,0.6}$	
No.	h_w [mm]	t_w [mm]	$k_G = 36$	$k_G = 31.6$	$k_G = 68.4$	$k_G = 59.2$	$k_G = 68.4$	$k_G = 59.2$
[1]	[2]	[3]	[4]	[5]	[6]	[7]	[8]	[9]
1	1000	6	1.12	1.13	0.68	0.69	0.83	0.85
2	1000	8	1.08	1.09	0.66	0.67	0.85	0.87
3	1000	10	1.04	1.05	0.64	0.65	0.85	0.87
4	1000	12	0.99	1.01	0.61	0.63	0.84	0.87
5	1000	14	0.95	0.97	0.59	0.60	0.82	0.86
6	1000	16	0.90	0.92	0.56	0.57	0.80	0.84
7	1000	18	0.86	0.89	0.54	0.55	0.78	0.82
8	1200	6	1.11	1.12	0.67	0.67	0.85	0.87
9	1200	8	1.08	1.09	0.65	0.66	0.86	0.89
10	1200	10	1.03	1.05	0.63	0.64	0.87	0.90
11	1200	12	0.99	1.01	0.60	0.62	0.86	0.90
12	1200	14	0.97	1.00	0.58	0.60	0.84	0.88
13	1200	16	0.93	0.96	0.55	0.57	0.83	0.87
14	1200	18	0.89	0.92	0.54	0.56	0.81	0.86
15	1400	6	1.13	1.14	0.67	0.68	0.88	0.90
16	1400	8	1.10	1.12	0.66	0.67	0.90	0.94
17	1400	10	1.05	1.08	0.64	0.65	0.91	0.95
18	1400	12	1.01	1.04	0.61	0.63	0.90	0.94
19	1400	14	0.98	1.02	0.59	0.61	0.89	0.93
20	1400	16	0.95	1.00	0.57	0.60	0.88	0.93
21	1400	18	0.94	0.98	0.56	0.59	0.87	0.92
22	1600	6	1.16	1.18	0.69	0.70	0.93	0.96
23	1600	8	1.13	1.16	0.67	0.69	0.96	1.00
24	1600	10	1.08	1.12	0.66	0.68	0.97	1.01
25	1600	12	1.05	1.09	0.63	0.66	0.96	1.01
26	1600	14	1.03	1.08	0.62	0.65	0.95	1.01
27	1600	16	1.02	1.07	0.61	0.64	0.95	1.01
28	1600	18	1.01	1.07	0.61	0.65	0.96	1.02
29	1800	6	1.18	1.21	0.70	0.71	0.97	1.01
30	1800	8	1.17	1.20	0.70	0.73	1.03	1.07
31	1800	10	1.13	1.17	0.68	0.71	1.03	1.08
32	1800	12	1.11	1.16	0.66	0.70	1.02	1.08
33	1800	14	1.10	1.16	0.65	0.69	1.02	1.09
34	1800	16	1.09	1.16	0.64	0.68	1.01	1.08
35	1800	18	1.09	1.16	0.63	0.68	1.01	1.08
36	2000	6	1.21	1.24	0.72	0.74	1.02	1.07
37	2000	8	1.21	1.25	0.73	0.76	1.08	1.14
38	2000	10	1.19	1.24	0.71	0.75	1.09	1.15
39	2000	12	1.18	1.24	0.70	0.74	1.09	1.16
40	2000	14	1.17	1.25	0.69	0.73	1.08	1.16
41	2000	16	1.17	1.25	0.68	0.73	1.08	1.15

(*continued on next page*)

Table 3.6 Comparison between the critical shear stress ($\tau_{cr,FE}$) of the CWs of simple and fixed juncture with first-order interactive buckling strength [3]—cont'd

No.	h_w [mm]	t_w [mm]	S		F		F	
			$\tau_{cr,FE}/\tau_{cr,I,1}$		$\tau_{cr,FE}/\tau_{cr,I,1}$		$\tau_{cr,FE}/\tau_{cr,I,0.6}$	
			$k_G = 36$	$k_G = 31.6$	$k_G = 68.4$	$k_G = 59.2$	$k_G = 68.4$	$k_G = 59.2$
42	2000	18	1.18	1.26	0.68	0.73	1.07	1.16
43	2200	6	1.25	1.29	0.73	0.76	1.07	1.11
44	2200	8	1.26	1.31	0.76	0.79	1.14	1.21
45	2200	10	1.25	1.32	0.74	0.78	1.16	1.23
46	2200	12	1.25	1.32	0.73	0.78	1.15	1.23
47	2200	14	1.25	1.33	0.72	0.77	1.14	1.23
48	2200	16	1.25	1.35	0.72	0.78	1.14	1.23
49	2200	18	1.27	1.37	0.72	0.78	1.14	1.24
50	2400	6	1.31	1.36	0.76	0.78	1.12	1.18
51	2400	8	1.33	1.39	0.79	0.82	1.21	1.28
52	2400	10	1.32	1.40	0.77	0.82	1.21	1.29
53	2400	12	1.32	1.40	0.76	0.81	1.20	1.29
54	2400	14	1.32	1.42	0.76	0.82	1.20	1.30
55	2400	16	1.34	1.45	0.76	0.83	1.21	1.31
56	2400	18	1.37	1.49	0.77	0.85	1.22	1.32
		Ave	**1.12**	**1.17**	**0.67**	**0.70**	**0.99**	**1.05**
		COV	**0.130**	**0.149**	**0.066**	**0.076**	**0.131**	**0.150**

buckling strength ($\tau_{cr,FE}/\tau_{cr,I,2}$) according to Abbas et al. [21] was found to be 0.53 and 0.54 by using $k_G = 36$ [19] and 31.6 [10], respectively.

Hence, the interactive buckling strength ($\tau_{cr,I}$) should be modified for this later case. For that reason, the authors found that the exponent (n) that fits with the *FE* results is 0.6, as given in Eq. (3.13). Accordingly, the authors recommend to calculate the interactive buckling strength using k_G of 59.2 [10] which is conservative than the other value; see the average values of Table 3.6.

The relationship between the relative *FE* buckling stress of simple junctures to first-order interactive buckling strength ($\tau_{cr,FE}/\tau_{cr,I,1}$) is presented in Figs 3.8 and 3.9 versus the h_r/t_w ratio. The difference between the curves of Figs 3.8 and 3.9 is a result of using different values for the global shear buckling coefficient (k_G) of 36 [19] and 31.6 [10], respectively. Confirming with Yi et al. [3], it can be seen that according to both Figs 3.8 and 3.9, the trapezoidally corrugated steel web plates need to satisfy the following geometric condition using k_G of 36 [19] *or* 31.6 [10]; $b/h_w \leq 0.2$.

On the other hand, Fig. 3.10 provides the same relationships but by using the modified interactive buckling strength ($\tau_{cr,FE}/\tau_{cr,I,0.6}$) for the case of CWs with fixed junctures. As it can be noticed, the trapezoidally corrugated steel web plates with fixed junctures need as well to satisfy the same geometric condition of $b/h_w \leq 0.2$. Accordingly, the web depth (h_w) of further simulations for trapezoidally corrugated steel web plate segments using the same corrugation dimensions, should not be less

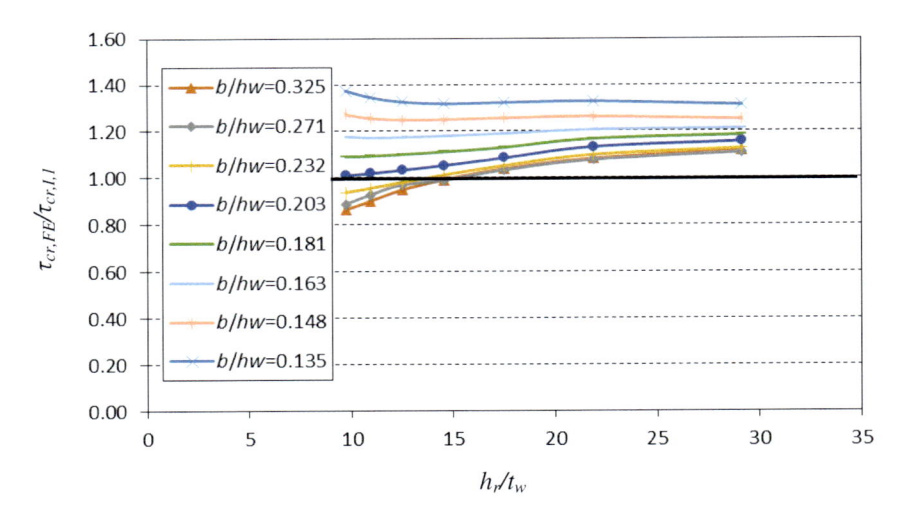

Figure 3.8 Relative finite element critical stress to first-order interactive buckling strength versus the h_r/t_w ratio (using k_G of 36 [19]).

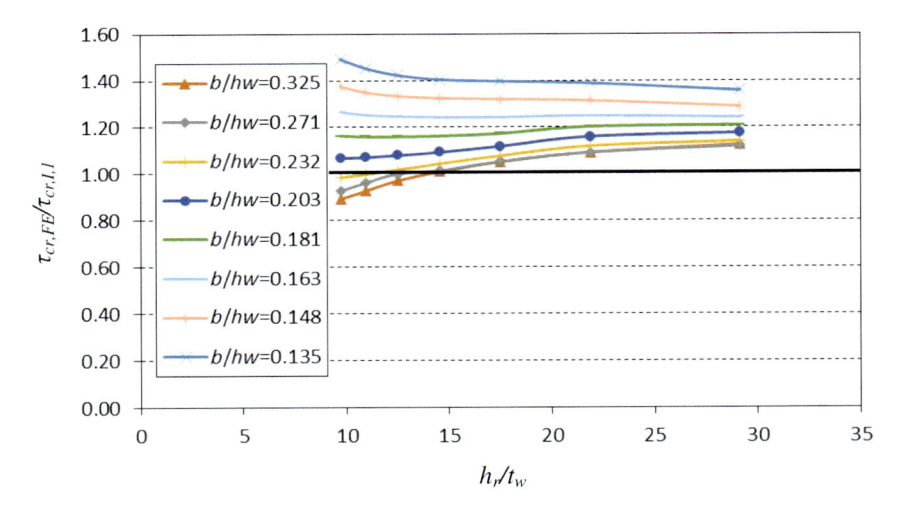

Figure 3.9 Relative finite element critical stress to first-order interactive buckling strength versus the h_r/t_w ratio (using k_G of 31.6 [10]).

than 1600 mm.

$$\tau_{cr,I,1} = \frac{\tau_{cr,L} \cdot \tau_{cr,G}}{\tau_{cr,L} + \tau_{cr,G}} \tag{3.12}$$

$$\tau_{cr,I,0.6} = \frac{\tau_{cr,L} \cdot \tau_{cr,G}}{\left(\left(\tau_{cr,L}\right)^{0.6} + \left(\tau_{cr,G}\right)^{0.6}\right)^{\frac{1}{0.6}}} \tag{3.13}$$

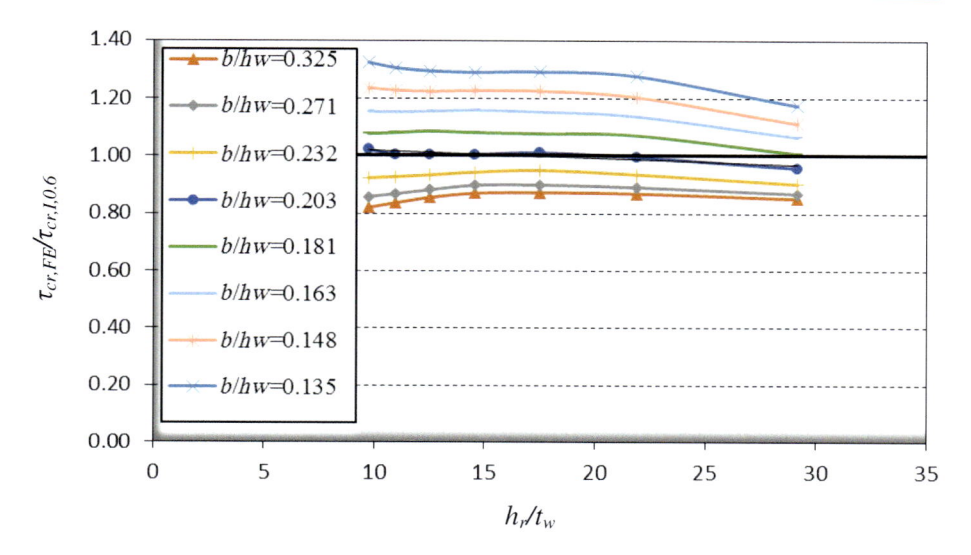

Figure 3.10 Relative finite element critical stress to first-order interactive buckling strength versus the h_r/t_w ratio (using k_G of 59.2 [10]).

3.8 Buckling mechanism

The generated FE models in the present investigation were designed merely to buckle by the shear of CW plates. As can be found in Section 3.2, steel CWs could develop three forms of shear buckling mechanisms; Local (L), Global (G), and Interactive (I). Examples for the buckling shapes, plotted from the visualization module of ABAQUS program [2] for the three buckling modes, are given in Figs 3.2–3.4. Only models of fixed junctures and web depth (h_w) $\geq$ 1600 mm were considered here. Generally, it could be concluded that there are no obvious limits that can be identified for the three shear buckling modes, which requires an extension to the current investigation in further research works.

As expected, the numerical results showed that the local buckling, as can be seen in Fig. 3.2, occurs when a flat sub-plate between vertical edges has a large width-to-thickness ratio. In the current modeling, the local buckling occurred only for corrugated plates of panel width-to-thickness ratio of 54 ($b = 325$ mm and $t_w = 6$ mm). Within the models that have a panel width-to-thickness ratio of 54, the local buckling was associated with the cases of $h_w/t_w \geq 250$. As can be seen in Fig. 3.2, local buckling involves single flat panels.

On the opposite, for the case of dense corrugations, global buckling involving multiple panels (see Fig. 3.3) becomes the governing buckling mode, with buckles extending diagonally over the whole depth of the web. It could be observed that the global buckling is affected by the combinations of h_r/t_w and h_w/t_w ratios. For example, decreasing h_r/t_w ratio (say for example 9.72) leads to global buckling at a low h_w/t_w ratio (100 for the case of F-1800-18) and vice versa.

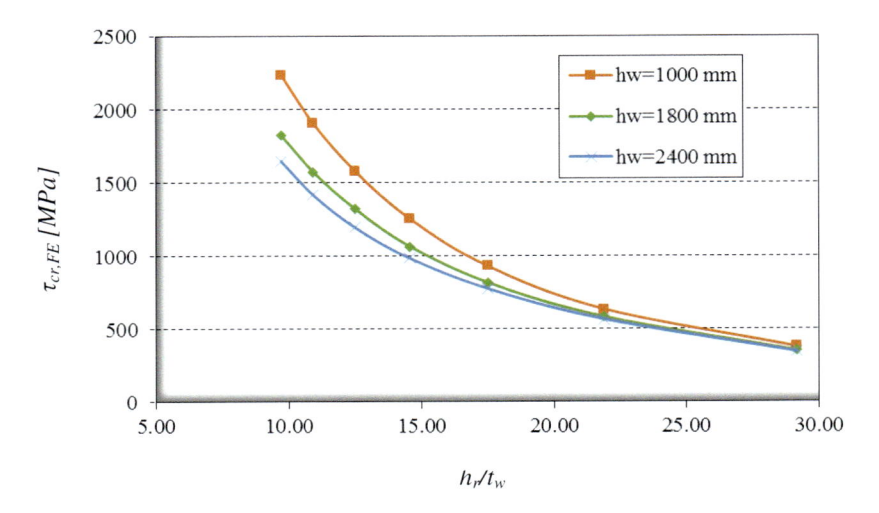

Figure 3.11 Finite element critical stress versus h_r/t_w ratio for sample girders.

Among the local and global buckling modes, the interactive buckling mode, which is not as definitive as those of the previously mentioned buckling modes, was additionally observed. It could be noticed that the interactive buckling is affected by the combinations of h_r/t_w and b/h_w ratios, as observed also by Yi et al. [3]. In the current models, the interactive buckling was observed mainly for the plates of the large b/h_w ratios with different h_r/t_w ratios; $b/h_w = 0.203$ ratios.

3.9 Effects of key parameters on plate segments

In this section, a structural analysis is made for other key parameters taken into consideration in the current research, as given in Section 3.4. Note that each parameter under this title is examined while other are kept fixed. In the following, the structural behavior of the trapezoidally corrugated steel web plates of the current FE program with *fixed* junctures is presented.

3.9.1 Effect of corrugation depth-to-web thickness ratio (h_r/t_w)

In the current FE program, seven corrugation depth-to-web thickness ratios (h_r/t_w) were examined. The ratios of h_r/t_w ranged from 10 to 30. The lower bound of 10 was recommended by Easley [19] to satisfy the required geometric stiffness of the CWs. The relationships between FE critical shear stress ($\tau_{cr,FE}$) and h_r/t_w ratios for three web depths (h_w) are given in Fig. 3.11. It can, however, be observed that increasing the h_r/t_w ratio reduces consequently the critical shear stress ($\tau_{cr,FE}$) for any web depth (h_w). In addition, it can be noticed that decreasing the web depth (h_w) results in a considerably higher critical shear stress ($\tau_{cr,FE}$) near the lower bound of h_r/t_w; 10. Furthermore, it is

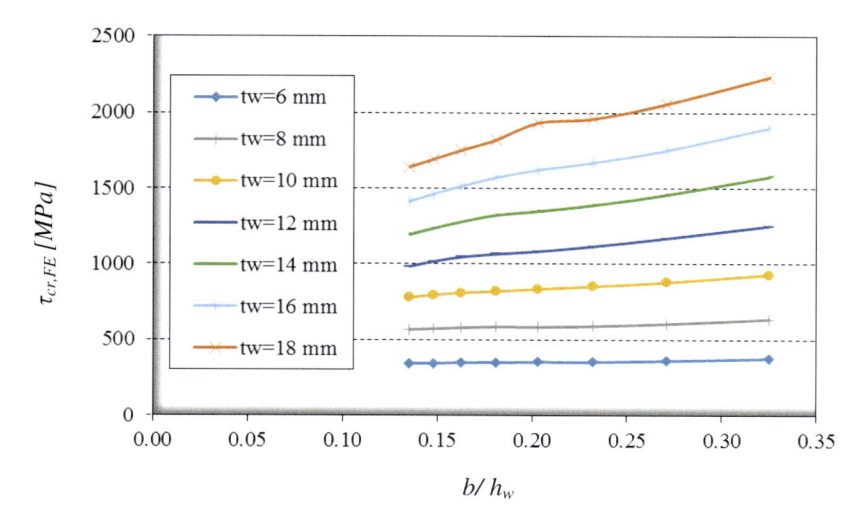

Figure 3.12 Finite element critical stress versus b/h_w ratio.

found that when the h_r/t_w ratio is close to 30, the critical shear stress ($\tau_{cr,FE}$) for any considered web depth (h_w) becomes nearly the same.

3.9.2 Effect of web flat panel width-to-depth ratio (b/h_w)

Only one flat plate width ($b = 325$ mm) was used in the current modeling. Hence, to evaluate the effect of the flat panel width-to-depth ratio (b/h_w), eight ratios of b/h_w were used in the current FE modeling program by changing the web depth (h_w) from 1000 to 2400 mm. Fig. 3.12 plots the relationship between the FE critical shear stress ($\tau_{cr,FE}$) and the b/h_w ratio, from which it can be noticed that increasing the b/h_w ratio (by reducing h_w) raises the critical shear stress ($\tau_{cr,FE}$) of the CW *linearly*. It should be noted that the previously mentioned increase in the critical shear stress ($\tau_{cr,FE}$) becomes larger as the CW thickness (t_w) increases.

3.9.3 Effect of aspect ratio of the web panel (a/h_w)

The effect of the aspect ratio of the web panel (a/h_w) for the trapezoidally corrugated steel web plates was additionally investigated. The a/h_w ratio varied from 1.0 to 2.4 by changing h_w from 1.0 m to 2.4 m for the same panel width (a) of 2.4 m. The average critical shear stress ($\tau_{cr,FE}$) was calculated for each panel width (a) taking into account the a/h_w ratio, as can be seen in Fig. 3.13. From this figure, it can certainly be seen that increasing the a/h_w ratio (of the same a) for any CW increases the critical shear stress ($\tau_{cr,FE}$). This indicates that square CWs are the most critical. However, this case does not comply with the conditions of CWGs used in practice, where the distance between the vertical stiffeners is much greater than the web depth (h_w) [15].

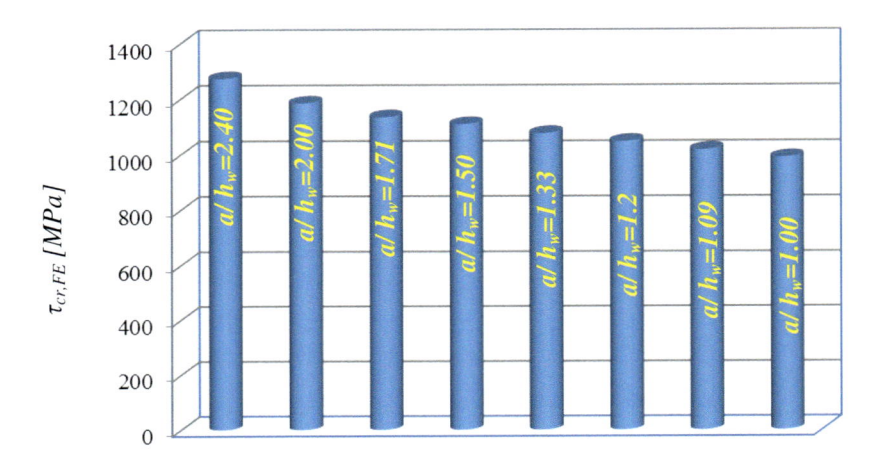

Figure 3.13 Average finite element critical stress for each a/h_w ratio.

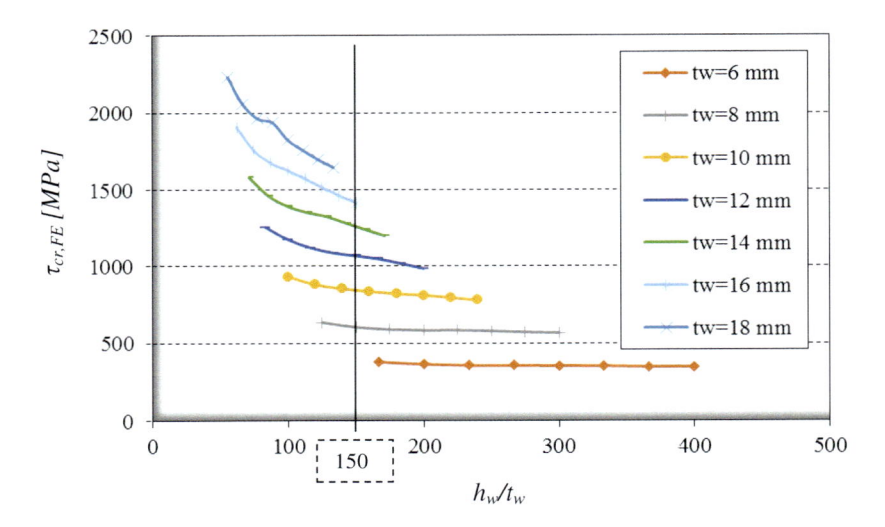

Figure 3.14 Finite element critical stress versus the h_w/t_w ratio.

3.9.4 Effect of web plate slenderness (h_w/t_w)

The effect of the CW plate slenderness (h_w/t_w) is discussed in this section. Several h_w/t_w ratios ranging from 56 to 400 have been examined in the current FE modeling program. The FE critical stress ($\tau_{cr,FE}$) versus the (h_w/t_w) ratio is shown in Fig. 3.14. It can be seen that increasing the h_w/t_w ratio subsequently reduces the critical shear stress ($\tau_{cr,FE}$) for any web thickness (t_w). It can also be observed that increasing the web thickness (t_w) leads to a significant decrease in the critical shear stress ($\tau_{cr,FE}$). On the opposite, the critical shear stress ($\tau_{cr,FE}$) for any considered web

Table 3.7 Sample results for the effect of flange width (b_f).

h_w [mm]	t_w [mm]	t_f [mm]	b_f [mm]	$\tau_{cr,FE}$ [MPa]
1000	6	50	200	377.00
			250	377.78
			300	378.37
			350	378.62
			400	378.93
			450	379.00
			500	379.43

depth (h_w) remains almost the same if the web thickness (t_w) is significantly reduced; see for example $t_w = 6$ mm. In addition, it is worth pointing out that when the h_w/t_w ratio exceeds 150, the critical stress ($\tau_{cr,FE}$) becomes *nearly constant* or its reduction becomes very small for each web thickness (t_w).

3.10 Real behavior between the CW and flanges

In order to evaluate the real juncture behavior between the CW and flanges, girder segments composed of trapezoidally corrugated steel web plates and flanges were additionally generated. Firstly, the effect of flange width (b_f) on the juncture behavior was investigated. After that, the new models of girder segments were developed. The critical shear stresses ($\tau_{cr,FE}$) of the newly generated segments were then compared with the corresponding values, given in Table 3.5, for the cases of simple and fixed CW plates.

3.10.1 Effect of flange width (b$_f$)

In order to investigate the effect of flange width (b_f), elastic bifurcation buckling analyses for girder segments with different flange widths (b_f) were performed, while keeping the flange thickness (t_f) constant. Sample results are provided in Table 3.7. From the presented results, the critical shear stress ($\tau_{cr,FE}$) for the girder segment of $b_f = 500$ mm over that of $b_f = 200$ mm is 1.006. Accordingly, the flange width (b_f) generally has limited effect on real behavior. For the extended parametric study on girder segments, a value of $b_f = 400$ mm was used.

3.10.2 Extended parametric study on girder segments

New models of girder segments were developed, using flanges of $b_f = 400$ mm. The current girder segments had the same web dimensions as those previously modeled in Section 3.4. The current study's recommendation (Section 3.7) that the b/h_w ratio should be less than and including 0.2 for both simple (S) and fixed (F) boundary conditions was used in the new generated models. Accordingly, the web depth (h_w) for further simulations of trapezoidally corrugated steel web plate segments using the same

Table 3.8 Comparison between girder segments with flanges and other plates with simple (S) and fixed (F) boundary conditions.

t_w [mm]	t_f/t_w	Relative critical stresses									
		$h_w = 1.6$ m		$h_w = 1.8$ m		$h_w = 2.0$ m		$h_w = 2.2$ m		$h_w = 2.4$ m	
		RF/S	RF/F	RF/S	RF/F	RF/S	RF/F	RF/S	RF/F	RF/S	RF/F
6	3	1.00	0.99	1.01	0.99	1.01	0.99	1.00	0.99	1.01	0.99
6	4	1.00	0.99	1.02	1.01	1.01	0.99	1.00	0.99	1.01	0.99
6	5	1.00	0.99	1.02	1.01	1.01	0.99	1.03	1.01	1.01	1.00
8	3	1.02	1.00	1.03	0.99	1.05	1.00	1.04	1.00	1.04	1.01
8	4	1.02	1.00	1.04	1.00	1.05	1.01	1.05	1.01	1.05	1.02
8	5	1.03	1.01	1.05	1.01	1.05	1.01	1.05	1.01	1.06	1.02
10	3	1.04	1.00	1.04	1.00	1.04	1.00	1.04	1.00	1.04	1.01
10	4	1.05	1.00	1.05	1.01	1.06	1.01	1.06	1.02	1.06	1.02
10	5	1.06	1.01	1.07	1.02	1.07	1.02	1.07	1.03	1.07	1.04
12	3	1.05	1.00	1.04	1.00	1.05	1.00	1.03	1.00	1.03	1.01
12	4	1.06	1.01	1.06	1.01	1.06	1.02	1.05	1.02	1.05	1.02
12	5	1.07	1.02	1.07	1.02	1.07	1.03	1.07	1.03	1.06	1.03
14	3	1.04	1.00	1.04	1.00	1.03	1.00	1.02	0.99	1.03	1.00
14	4	1.06	1.01	1.06	1.01	1.05	1.02	1.04	1.01	1.04	1.01
14	5	1.07	1.02	1.07	1.02	1.07	1.03	1.05	1.02	1.05	1.03
16	3	1.05	1.00	1.04	1.00	1.02	0.99	1.03	1.00	1.02	1.00
16	4	1.06	1.01	1.06	1.02	1.04	1.01	1.04	1.01	1.03	1.01
16	5	1.07	1.02	1.07	1.03	1.05	1.02	1.05	1.02	1.04	1.02
18	3	1.05	0.98	1.03	0.99	1.03	1.00	1.02	0.99	1.03	1.01
18	4	1.06	0.99	1.04	1.01	1.04	1.01	1.03	1.01	1.04	1.01
18	5	1.07	1.00	1.06	1.02	1.05	1.02	1.04	1.02	1.05	1.02
Ave		**1.04**	**1.00**	**1.05**	**1.01**	**1.04**	**1.01**	**1.04**	**1.01**	**1.04**	**1.01**
COV		**0.024**	**0.012**	**0.018**	**0.011**	**0.020**	**0.014**	**0.018**	**0.012**	**0.019**	**0.013**

corrugation dimensions was equal to or greater than 1600 mm. The flange thickness (t_f) was proposed assuming t_f/t_w ratios greater than and including 1.0.

3.10.3 Results and evaluation

The results clearly indicated that when the flanges are sufficiently rigid ($t_f/t_w \geq 3.0$), the girder segments exhibit shear buckling mechanisms. Whereas if the CW plates are relatively rigid ($t_f/t_w < 3.0$), the buckling of the girders becomes controlled by the deformation of the flanges. The critical shear stresses ($\tau_{cr,FE,\ RF}$) for the models of $t_f/t_w \geq 3.0$ are obtained from the FE modeling results. They are then compared with the stress values of the simple (S) and fixed (F) boundary conditions plates mentioned earlier, as can be found in Table 3.8. From the average ratios, it can be clearly concluded that the real juncture is nearly fixed. Hence, the available interactive shear buckling strength formulas, as can be found here in Table 3.2, *cannot be adequately* used. Instead, the proposed formula using ($n = 0.6$) as proposed in Eq. 3.13 should be used.

References

[1] S.C. Lee, J.S. Davidson, C.H. Yoo, Shear buckling coefficients of plate girder web panels, Comput. Struct. 59 (1996) 789–795.

[2] ABAQUSABAQUS Standard User's Manual, 2008 The Abaqus Software is a product of Dassault Systèmes Simulia Corp., Providence, RI, USA Dassault Systèmes, Version 6.8.

[3] J. Yi, H. Gil, K. Youm, H. Lee, Interactive shear buckling behavior of trapezoidally corrugated steel webs, Eng. Struct. 30 (2008) 1659–1666.

[4] J. Moon, J. Yi, B.H. Choi, H. Lee, Shear strength and design of trapezoidally corrugated steel webs, J. Constr. Steel Res. 65 (2009) 1198–1205.

[5] M.E.A. Eldib, Shear buckling strength and design of curved corrugated steel webs for bridges, J. Constr. Steel Res. 65 (2009) 2129–2139.

[6] R.W. Hamilton, "Behavior of welded girder with corrugated webs", Ph.D. thesis, University of Maine, Orono, Maine, 1993.

[7] R.G. Driver, H.H. Abbas, R. Sause, Shear behavior of corrugated web bridge girders, J. Struct. Eng., ASCE 132 (2) (2006) 195–203.

[8] J.T. Easley, D.E. McFarland, Buckling of light-gage corrugated metal shear diaphragms, J. Struct. Div., ASCE 95 (1969) 1497–1516.

[9] J. Lindner, R. Aschinger, Grenzschubtragfähigkeit von I-trägern mit trapezförmig profilierten Stegen, Stahlbau 57 (12) (1988) 377–380.

[10] M. Elgaaly, R.W. Hamilton, A. Seshadri, Shear strength of beams with corrugated webs, J. Struct. Eng. 122 (4) (1996) 390–398.

[11] A.S. El-Metwally Prestressed composite girders with corrugated steel webs. M.S. thesis, Calgary (AB), Department of Civil Engineering, University of Calgary; 1998.

[12] H.H. Abbas, "Analysis and design of corrugated web I-girders for bridges using high performance steel", Ph.D. dissertation, Bethlehem, PA, Department of Civil and Environmental Engineering, Lehigh University, 2003.

[13] R. Sause, H.H. Abbas, W. Wassef, R.G. Driver, M. Elgaaly, Corrugated web girder shape and strength criteria: work area 1. Pennsylvania innovative high performance steel bridge demonstration project, 2003 ATLSS report no. 03-18. Bethlehem, PA, Advanced Technology for Large Structural Systems. ATLSS. Center. Lehigh University.

[14] R.G. Driver, H.H. Abbas, R. Sause, Shear behavior of corrugated web bridge girders, J. Struct. Eng., ASCE 132 (2) (2006) 195–203.

[15] R. Sause, T.N. Braxtan, Shear strength of trapezoidal corrugated steel webs, J. Constr. Steel Res. 67 (2011) 223–236.

[16] S.P. Timoshenko, J.M. Gere, Theory of Elastic Stability, 2nd ed., McGraw-Hill Publishing co., New York, NY, 1961.

[17] P.S. Bulson, Stability of Flat Plates, Elsevier, New York, NY, 1970.

[18] T.V. Galambos, Guide to Stability Design Criteria for Metal Structures, John Wiley & Sons, Inc., New York, NY, 1988.

[19] J.T. Easley, Buckling formulas for corrugated metal shear diaphragms, J. Struct. Div., SECF ST7 (1975) 1403–1417.

[20] A. Bergfelt, L. Leiva "Shear buckling of trapezoidally corrugated girders webs", Report part 2. Pibl.SS4:2. Sweden, Chalmers University of Technology; 1984.

[21] H.H. Abbas, R. Sause, RG. Driver, Shear strength and stability of high performance steel corrugated web girders, in: SSRC conference, 2002, pp. 361–387.

[22] H. Shiratoni, H. Ikeda, Y. Imai, K. Kano, Flexural shear behavior of composite bridge girder with corrugated steel webs around middle support, JSCE J. 724 (I-62) (2003) 49–67.

[23] E.Y. Sayed-Ahmed, Plate girders with corrugated steel webs, AISC Eng. J., First Quarter (2005) 1–13.

[24] R. Luo, B. Edlund, Ultimate strength of girders with trapezoidally corrugated webs under patch loading, Thin-Walled Struct. 24 (1996) 135–156.

Shear buckling behavior

4

4.1 Scope

Until the middle of the 1990s, the shear buckling mechanism in steel corrugated webs was classified only as *local* and *global* buckling. In the local buckling mode, the corrugated web acts as a series of flat plate subpanels that mutually support each other along their vertical edges and are supported by the flanges at their horizontal edges. The global buckling mode involves multiple folds and the buckled shape extends diagonally over the depth of the web. However, it was noticed for example by Hamilton [1] that *the failure in some specimens was initiated by local buckling of one of the corrugation folds and then propagated to other folds*. This mode of failure was classified and dealt with in the literature [1,2] as a local buckling mode. Later on, this mode of failure which has the characteristics of both *local* and *global* buckling modes was classified as the *interactive* buckling mode supported by extensive experimental observations.

For a steel girder with corrugated web, it is assumed that the web purely carries shear forces due to the accordion effect [1–7]. Corrugated web does not carry significant longitudinal stresses from the primary flexure of the girders and, consequently, the bending moment can reasonably be assumed to be carried totally by the flanges. Recently, Moon et al. [8] investigated the lateral-torsional buckling of I-girder with corrugated webs under uniform bending. The flexural strength of such girders was found to be carried only by the flanges. It is worth pointing out that, for that reason, the shear strength can be determined without consideration of moment-shear interaction. Accordingly, the shear stress of the web can be assumed constant and can be calculated using the following average shear stress:

$$\tau = \frac{V}{t_w h_w} \tag{4.1}$$

On the other hand, the problem of shear buckling of bridge girders has been the focus of a significant amount of studies in recent years. As a result, a significant amount of research has been carried out to propose shear design formula for bridge girders with corrugated webs (BGCWs) [3–7]. It was, however, found that *the shear strength formula used in design are conservative* compared to the test data [7]. This was attributed to the very little data on (1) the actual initial imperfection of corrugated web folds or on other imperfections in corrugation geometry, (2) the effect of bend radius at the fold lines of corrugated webs, and (3) the effects of residual stresses on shear strength. In addition, the current authors noticed that, for practical design purposes, lower bound values for local (k_L) and global (k_G) buckling factors are generally used in the available formula. This means that the case of simply supported edges with the flanges for the longitudinal/inclined panels and the entire corrugated webs,

Behavior and Design of Trapezoidally Corrugated Web Girders for Bridge Construction: Recent Advances.
DOI: https://doi.org/10.1016/B978-0-323-88437-2.00004-6

respectively, are considered. However, this assumption was found by the current authors [9] to be *conservative*, as it was concluded that the realistic support condition at the juncture is nearly fixed for $t_f/t_w \geq 3.0$; refer to the previous chapter. Accordingly, from the authors' viewpoint *this might increase the margin of safety* taken into consideration in the current shear buckling designs of BGCWs. Hence, the work by Hassanein and Kharoob [9] was extended to suggest a design shear strength formula that better fits the available experimental results for full-scale BGCWs as well as finite element (FE) modeling results.

4.2 Effect of initial imperfection

The amplitude and shape of the initial web geometric imperfections were found to play a significant role in the shear strength and behavior of steel girders with corrugated webs [2,3,10]. For instance, from FE simulations performed by Elgaaly et al. [2] to study the behavior of steel beams with corrugated webs under shear, it was found that the analytical results are higher than the corresponding experimental ones. This discrepancy, as concluded by Elgaaly et al. [2], was attributed to the presence of unavoidable out-of-plane initial imperfections in the test specimens. In addition, Moon et al. [10] measured the initial imperfection amplitude in the experimental tests conducted to investigate the shear behavior and strength of corrugated steel webs. They found that the maximum magnitude of the initial imperfection is 17.9 mm. However, it was mentioned [10] that while the tests were carried out using large-scale specimens, the web thickness of the test specimen (4 mm) was relatively smaller than that used in actual bridges. Moon et al. [10] found that this is one of the reasons for the large initial imperfection of the study, and the magnitude of the initial imperfections in actual bridges may be smaller than that of their study. Accordingly, *they concluded that further study for the initial imperfections of real bridges with corrugated webs should be made for the rational application of their design proposed equation.* Furthermore, the results of FE analyses conducted by Driver et al. [3] suggested that the strength is overestimated despite the lower bound assumption of simply supported edges at the web panel boundaries, at least in part, because of the sensitivity of the shear behavior to the presence of initial imperfections in the web. Accordingly, Driver et al. [3] studied the effect of web initial geometric imperfections on the shear strength. From these studies, it was first found that, given the size of the panels in the webs of available bridge girders, *an imperfection amplitude equalling to the corrugated web thickness (t_w) is realistic.* Taking into account the web imperfection shape in their second study [3], it was concluded that the shear stress capacity generally increases with the mode number and *the first mode provides the most critical condition.*

4.3 Normal-strength steel prismatic girders

In this section, emphasis is placed on prismatic girders made of normal-strength steel (NSS).

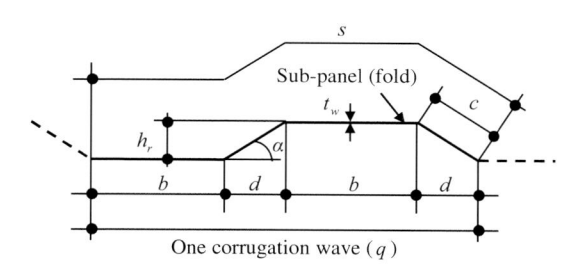

Figure 4.1 Corrugation configuration and geometric notation.

Table 4.1 Profiles of available test for bridge girders with corrugated webs [3,10–11].

Girder	b [mm]	d [mm]	t_w [mm]	h_w [mm]	α [°]	τ_e/τ_y
G7A [3]	300	200	6.3	1500	36.9	0.91
G8A [3]	300	200	6.27	1500	36.9	0.85
M12 [9]	250	220	4.0	2000	17.18	0.64
M13 [9]	220	180	4.0	2000	14.63	0.62
M14 [9]	220	180	4.0	2000	18.72	0.77
L1 [12]	450	300	4.8	1500	33.7	0.72
L2 [12]	550	300	4.8	1500	32.2	0.60
L3 [12]	450	300	4.8	1500	9.4	0.51
L4 [12]	550	300	4.8	1500	10.6	0.46
I1 [12]	320	100	4.8	2000	24	0.95
I2 [12]	350	100	3.8	2000	16	0.52
G1 [12]	200	180	4.8	2000	14.2	0.79
G2 [12]	160	50	3.8	2000	33.4	0.83
G3 [12]	160	100	3.8	2000	15.1	0.85

4.3.1 Comparison between design shear strength formula and available experimental full-scale tests

The normalized experimental shear strengths (τ_e/τ_y) of the full-scale BGCWs conducted by Driver et al. [3], Moon et al. [10], and Gil et al. [11] are herein compared to the available design formula suggested by Driver et al. [3], Sause and Braxtan [7], and Moon et al. [10]. A brief description for these shear design formula is provided in the following paragraphs, using the geometric notations as shown in Fig. 4.1.

4.3.1.1 Full-scale tests

The results of five full-scale BGCWs having $t_f/t_w \geq 3.0$, which were tested experimentally by Moon et al. [10] and Driver et al. [3], are used herein to check the validity of available design shear strengths of Driver et al. [3], Sause and Braxtan [7], and Moon et al. [10]. The corrugation details of the tests are given in Table 4.1. Additionally, another nine bridge girders tested by Gil et al. [11] are used to check the validity of the available design shear strengths. It should be mentioned that the details of these experiments were reported by Sause and Braxtan [7] without the details of the flanges of the bridges. However, they were checked herein expecting that $t_f/t_w \geq 3.0$.

4.3.1.2 Design strengths in the literature

4.3.1.2.1 Design shear buckling strength ($\tau_{n,M}$) by Moon et al. [10]

In order to calculate the shear buckling strength ($\tau_{n,M}$) by the method suggested by Moon et al. [10], the shear buckling parameter of corrugated webs (λ_s) should first be calculated by Eq. (4.2). Then, the interactive shear buckling coefficient (k_I) is to be calculated as defined by Eq. (4.3). This equation takes into consideration the lower bound values of the local (k_L) and global (k_G) buckling factors; *5.34 and 36, respectively*. The shear strength of corrugated webs can then be determined directly using Eq. (4.4), which adopts the buckling curve from the design manual for PC bridges with corrugated steel webs of Japan Society of Civil Engineers (JSCE) [13]. It is clear that calculating the shear buckling strengths according to Moon et al. [10] does not require calculating the local and global shear buckling strengths. This shear buckling strength ($\tau_{cr,M}$) is based upon the first-order interactive buckling strength proposed by Yi et al. [4].

$$\lambda_s = 1.05\sqrt{\frac{\tau_y}{k_I E}}\left(\frac{h_w}{t_w}\right) \tag{4.2}$$

$$k_I = \frac{30.54}{5.34(h_r/t_w)^{-1.5} + 5.72(w/h_w)^2} \tag{4.3}$$

$$\frac{\tau_{n,M}}{\tau_y} = \begin{cases} 1.0 & : \lambda_s \leq 0.6 \\ 1 - 0.614(\lambda_s - 0.6) & : 0.6 < \lambda_s \leq \sqrt{2} \\ \dfrac{1}{\lambda_s^2} & : \sqrt{2} < \lambda_s \end{cases} \tag{4.4}$$

4.3.1.2.2 Design shear buckling strength ($\tau_{n,D}$) according to Driver et al. [3]

Driver et al. [3] suggested to use Eq. (4.5) to obtain the nominal shear strength ($\tau_{n,D}$) of BGCWs. This equation takes into account the influences of both local and global buckling modes in a single interaction formula. Eq. (4.5), as suggested, could be applied over the full range of behavior, including cases where inelastic buckling and yielding controls.

$$\tau_{n,D} = \sqrt{\frac{\left(\tau_{cr,L}\cdot\tau_{cr,G}\right)^2}{\left(\tau_{cr,L}\right)^2 + \left(\tau_{cr,G}\right)^2}} \tag{4.5}$$

The elastic shear buckling stress is provided herein Eq. (4.3), while the global shear buckling stress, as adopted from the PhD thesis of Abbas [14], is given by Eq. (4.6):

$$\tau_{cr,G} = k_G F(\alpha, \beta)\frac{E t_w^{0.5} b^{1.5}}{12 h_w^2} \tag{4.6}$$

where k_G is the global buckling coefficient. Elgaaly et al. [2] assumed that the web is relatively long compared to h_w and suggested that k_G is to be taken as 31.6 and 59.2 assuming simple and fixed web-flange boundary conditions, respectively. Easely [15], however, suggested that k_G varies between 36 and 68.4. Consequently, the lower bound values for the local (k_L) and global (k_G) buckling factors according to Driver et al. [3] were recommended as *5.34 and 31.6, respectively*. $F(\alpha, \beta)$ is a coefficient based on the web corrugation geometry as follows:

$$F(\alpha, \beta) = \sqrt{\frac{(1 + \beta)\sin^3\alpha}{\beta + \cos\alpha} \cdot \left(\frac{3\beta + 1}{\beta^2(\beta + 1)}\right)^{0.75}} \tag{4.7}$$

where β is the ratio of b to c; and α is the corrugation angle. It should be noted that the values of $\tau_{cr,L}$ and $\tau_{cr,G}$ should not be greater than $0.8\tau_y$. Otherwise, inelastic shear buckling strength according to Elgaaly et al. [2] should be used.

4.3.1.2.3 Design shear buckling strength ($\tau_{n,S}$) according to Sause and Braxtan [7]

In 2011, Sause and Braxtan [7] proposed Eq. (4.8) for the design shear buckling strength ($\tau_{n,S}$).

$$\tau_{n,S} = \tau_y \left(\frac{1}{\left(\lambda_{I,3}\right)^6 + 2}\right)^{1/3} \tag{4.8}$$

where $\lambda_{I,3}$ is the interactive slenderness parameter which is calculated from Eq. (4.9) with $n = 3$. The local (λ_L) and global (λ_G) slenderness parameters were suggested to be determined from Eqs. 4.10 and 4.11. Note that the global (k_G) buckling factor calculated by Sause and Braxtan [7] using the suggestions of Easely [15]; k_G varies between 36 and 68.4 for simple and fixed boundary conditions, respectively. The global shear buckling stress was calculated using Eq. (4.6).

$$\lambda_{I,n} = \lambda_L \lambda_G \left(\left(\frac{1}{\lambda_L}\right)^{2n} + \left(\frac{1}{\lambda_G}\right)^{2n}\right)^{1/2n} \tag{4.9}$$

$$\lambda_L = \sqrt{\frac{12(1 - \upsilon^2)\tau_y}{k_L \pi^2 E} \frac{w}{t_w}} \tag{4.10}$$

$$\lambda_G = \sqrt{\frac{12h_w^2 \tau_y}{k_G F(\alpha, \beta)E t_w^{0.5} b^{1.5}}} \tag{4.11}$$

It is worth pointing out that this design strength ($\tau_{n,S}$) was verified by Sause and Braxtan [7] by considering a final set of 22 test specimen results. However, it was noticed by the current authors that the web depth (h_w) of the girders is ranging from 298 mm to 2000 mm with an average of 606 mm. This relatively small scale of the test specimens (particularly the web thickness (t_w) which was as small as 0.64 mm) almost had an influence on these shear test results. Considerable differences in material

Table 4.2 Comparison between available tests for bridge girders with corrugated webs [3,10,11] and design strengths.

Girder	τ_e/τ_y	$\tau_{n,M}/\tau_y$	$\tau_{n,D}/\tau_y$	$\tau_{n,S}/\tau_y$	$\tau_{n,M,0.6}/\tau_y$	$\tau_{n,D,0.6}/\tau_y$	$\tau_{n,S,0.6}/\tau_y$
G7A [3]	0.91	0.84	0.71	0.63	0.88	1.58	0.76
G8A [3]	0.85	0.83	0.71	0.63	0.88	1.57	0.76
M12 [9]	0.64	0.59	0.71	0.63	0.60	0.64	0.56
M13 [9]	0.62	0.62	0.65	0.61	0.65	0.72	0.60
M14 [9]	0.77	0.71	0.71	0.63	0.72	0.90	0.67
L1 [12]	0.72	0.67	0.63	0.59	0.78	1.08	0.71
L2 [12]	0.60	0.52	0.47	0.47	0.66	0.75	0.61
L3 [12]	0.51	0.58	0.63	0.59	0.63	0.69	0.59
L4 [12]	0.46	0.46	0.47	0.47	0.54	0.56	0.50
I1 [12]	0.95	0.63	0.71	0.63	0.68	0.79	0.63
I2 [12]	0.52	0.34	0.53	0.54	0.39	0.39	0.37
G1 [12]	0.79	0.75	0.69	0.63	0.77	1.07	0.71
G2 [12]	0.83	0.57	0.60	0.59	0.67	0.78	0.62
G3 [12]	0.85	0.46	0.53	0.54	0.56	0.58	0.52
Ave	0.72	0.61	0.62	0.58	0.67	0.86	0.62
COV	0.159	0.143	0.089	0.057	0.132	0.355	0.107

stress-strain behavior, and the geometric imperfections and residual stresses induced by fabricating the web and welding the web to the flanges, should be expected between such thin sheet material and the plate material used in actual bridge girders [3]. Therefore, this group was not used in the current comparison with design shear strength formula because it could not represent the behavior of actual bridges.

4.3.1.3   Comparison between available tests for bridge girders with corrugated webs and design strengths

The ratios of τ_e/τ_y of the full-scale BGCWs tested experimentally by Moon et al. [10], Driver et al. [3], and Gil et al. [11] are compared in this section with the design shear strengths $\tau_{n,M}$, $\tau_{n,D}$, and $\tau_{n,S}$, as can be found in Table 4.2. The same experimental ratios are compared with the design shear strengths $\tau_{n,M}$, $\tau_{n,D}$, and $\tau_{n,S}$ using the proposed interactive shear buckling strength formula ($\tau_{cr,I,0.6}$) presented in the previous chapter; $\tau_{n,M,0.6}$, $\tau_{n,D,0.6}$, and $\tau_{n,S,0.6}$, respectively. The following modifications were made to the previous shear design equations to calculate $\tau_{n,M,0.6}$, $\tau_{n,D,0.6}$, and $\tau_{n,S,0.6}$, respectively:

$$\lambda_{I,0.6} = \sqrt{\frac{\tau_y}{\tau_{cr,I,0.6}}} \tag{4.12}$$

$$\tau_{n,D,0.6} = \tau_{cr,I,0.6} \tag{4.13}$$

$$\tau_{n,S,0.6} = \tau_y \left(\frac{1}{\left(\lambda_{I,0.6}\right)^6 + 2}\right)^{1/3} \tag{4.14}$$

It can be noticed from Table 4.2 that, among the available design shear strengths, the design strength proposed by Moon et al. [10] and Driver et al. [3] are the most accurate. On the opposite, the design strength proposed by Sause and Braxtan [7] is the most conservative prediction. However, using the interactive shear buckling strength formula ($\tau_{cr,I,\,0.6}$) as proposed by the current authors [9] with the buckling curve included in the design manual for PC bridges with corrugated steel webs [13] is the most *suitable*. This may be attributed to the adoption of the upper bound values for the local (k_L) and global (k_G) buckling factors; *8.98 and 59.2, respectively*, because the realistic support condition at the juncture was found to be fixed for practical girders having $t_f/t_w \geq$ 3.0. The comparisons conducted herein, additionally, indicate that the interactive shear buckling strength formula ($\tau_{cr,I,\,0.6}$) proposed by the current author [9] *cannot be used* to predict the shear strength of BGCWs over the full range of behavior including yielding and inelastic domains.

4.3.2 Additional validation of design models

The conclusions presented in the previous section should further be verified putting in mind that the three girders tested by Moon et al. [10] and some of Gil et al. [11] *do not satisfy* the geometric condition of $\alpha \geq 30°$, as proposed by Lindner and Huang [16]. In addition, the test specimens failed only (except tests L4 and I2), as noticed from the comparison, by inelastic shear buckling; $0.6 < \lambda_s \leq \sqrt{2}$ in accordance with the buckling curve [13]. Also, adopting the interactive shear buckling strength formula ($\tau_{cr,I,\,0.6}$) as proposed by the current authors [9] with the buckling curve of the design manual for PC bridges with corrugated steel webs [13] assumes that it is possible for stocky corrugated webs to reach the yield shear strength ($\tau_n = \tau_y$). However, there is little test data that supports this result as concluded by Sause and Braxtan [7]. Hence, girders failing in the yielding domain should also be checked. Therefore, FE models for full-scale BGCWs were conducted to substitute the lack of knowledge of these kinds of girders and to verify the currently proposed shear strength in different behavioral domains [17].

The modeled full-scale BGCWs were loaded to fail under shear [17]. Accordingly, the failure was due to the web buckling. The modes of failure were developed by interactive (I) or global (G) shear buckling, as can be seen in Table 4.3. A sample of the load-mid-span deflection for the girders of $h_w = 1600$ mm can be seen in Fig. 4.2. As can be noticed, failure is sudden and results from buckling. The load-carrying capacity drops at the failure load and the specimen exhibits some residual strength after failure. In addition, it can be seen that the initial stiffness of the girder decreases with the decrease of its corrugated web thickness.

FE normalized shear strengths (τ_{FE}/τ_y) are compared herein with the previously mentioned design strengths, as can be seen in Table 4.3. It can be noticed that on the yielding and inelastic domains, the design strengths according to Driver et al. [3] and Sause and Braxtan [7] are *conservative*, while the strength by the method of Moon et al. [10] is mostly *unconservative* especially in the yielding zone. However, among the strengths using the proposed interactive shear buckling strength formula ($\tau_{cr,I,\,0.6}$) by the current authors [9], one that adopts the equation of Sause and Braxtan [7] is the

Table 4.3 Finite element results and comparison with design [3,7,10] and proposed strengths.

Girder (h_w-t_w)	Failure modes	$\dfrac{\tau_{FE}}{\tau_y}$	$\dfrac{\tau_{n,M}}{\tau_y}$	$\dfrac{\tau_{n,D}}{\tau_y}$	$\dfrac{\tau_{n,S}}{\tau_y}$	$\dfrac{\tau_{n,M,0.6}}{\tau_y}$	$\dfrac{\tau_{n,D,0.6}}{\tau_y}$	$\dfrac{\tau_{n,S,0.6}}{\tau_y}$	$\dfrac{\tau_{n,S,0.6}}{\tau_{FE}}$
1600-6	I	0.85	0.86	0.71	0.63	0.91	1.80	0.77	0.91
1600-8	I	0.92	1.00	0.71	0.63	1.00	2.86	0.79	0.85
1600-10	I	0.79	1.00	0.71	0.63	1.00	4.02	0.79	1.00
1600-12	G	0.85	1.00	0.71	0.63	1.00	5.27	0.79	0.93
1800-6	I	0.90	0.86	0.71	0.63	0.90	1.70	0.77	0.86
1800-8	I	0.91	0.97	0.71	0.63	0.99	2.66	0.79	0.87
1800-10	I	0.80	1.00	0.71	0.63	1.00	3.71	0.79	0.99
1800-12	G	1.01	1.00	0.71	0.63	1.00	4.81	0.79	0.78
2000-6	I	0.82	0.85	0.71	0.63	0.88	1.60	0.76	0.94
2000-8	I	0.93	0.96	0.71	0.63	0.98	2.48	0.79	0.84
2000-10	I	0.87	1.00	0.71	0.63	1.00	3.42	0.79	0.91
2000-12	G	0.83	1.00	0.71	0.63	1.00	4.40	0.79	0.96
2200-6	I	0.85	0.84	0.71	0.63	0.87	1.51	0.76	0.89
2200-8	I	0.87	0.95	0.71	0.63	0.96	2.32	0.78	0.90
2200-10	I	0.89	1.00	0.71	0.63	1.00	3.17	0.79	0.89
2200-12	G	0.67	1.00	0.71	0.63	1.00	4.04	0.79	1.19
2400-6	I	0.78	0.83	0.71	0.63	0.85	1.43	0.75	0.97
2400-8	I	0.79	0.94	0.71	0.63	0.95	2.17	0.78	0.99
2400-10	I	0.86	1.00	0.71	0.63	1.00	2.94	0.79	0.92
2400-12	G	0.72	1.00	0.71	0.63	1.00	3.72	0.79	1.09
	Ave	0.85	0.95	0.71	0.63	0.97	3.00	0.78	0.93
	COV	0.077	0.066	0.000	0.000	0.051	1.150	0.012	0.090

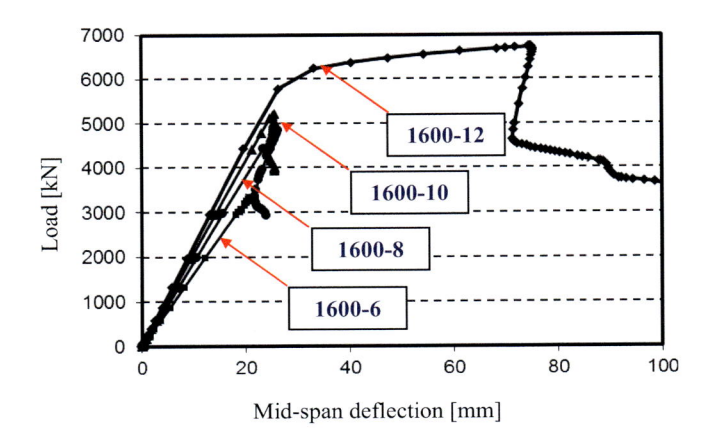

Figure 4.2 Load versus mid-span deflection for a sample of models.

Table 4.4 Comparison between interactive slenderness parameters $\lambda_{I,3}$ (Eq. 4.9) and $\lambda_{I,0.6}$ (Eq. 4.12).

Girder (h_w-t_w)	$\lambda_{I,3}$	$\lambda_{I,0.6}$	$\frac{\lambda_{I,0.6}}{\lambda_{I,3}}$
1600-6	1.12	0.75	0.66
1600-8	1.12	0.59	0.53
1600-10	1.12	0.50	0.44
1600-12	1.12	0.44	0.39
1800-6	1.12	0.77	0.68
1800-8	1.12	0.61	0.55
1800-10	1.12	0.52	0.46
1800-12	1.12	0.46	0.41
2000-6	1.12	0.79	0.70
2000-8	1.12	0.63	0.57
2000-10	1.12	0.54	0.48
2000-12	1.12	0.48	0.42
2200-6	1.12	0.81	0.72
2200-8	1.12	0.66	0.59
2200-10	1.12	0.56	0.50
2200-12	1.12	0.50	0.44
2400-6	1.12	0.84	0.75
2400-8	1.12	0.68	0.61
2400-10	1.12	0.58	0.52
2400-12	1.12	0.52	0.46
		Ave	0.54
		COV	0.112

best, as given in Eq. (4.14). Additionally, the current comparison points out that the interactive shear buckling strength formula ($\tau_{cr,I, 0.6}$) proposed by the current author [9], in line with that of Driver ae al [3], *cannot be used* to predict the shear strength of BGCWs over the full range of behavior.

The interactive slenderness parameters $\lambda_{I,3}$ and $\lambda_{I,0.6}$ from Eqs. (4.9) and (4.12), respectively, are provided in Table 4.4. It should be noted that the interactive slenderness parameter $\lambda_{I,0.6}$ is lower than $\lambda_{I,3}$ with an average ratio of $\lambda_{I,0.6}/\lambda_{I,3}$ of 0.54. As a result, the shear strengths of bridge girders using the real fixed juncture assumption [9] become higher than the corresponding values that use the practical design case of simple juncture. The normalized shear strength (τ_n/τ_y) versus the interactive shear buckling parameter (λ) is plotted for FE models as can be seen in Fig. 4.3. In this figure, the dotted vertical lines represent the yielding ($\lambda = 0.6$) and inelastic ($\lambda = \sqrt{2}$) shear buckling limits, respectively, according to the buckling curve of the design manual for PC bridges with corrugated steel webs [13]. Although Moon et al. [10] assume that the shear strength of stocky corrugated webs, ($\lambda \leq 0.6$), can possibly attain the shear yield stress of the material, FE results *do not support* this theoretical issue, at least in the range considered herein; $0.44 < \lambda \leq 0.6$.

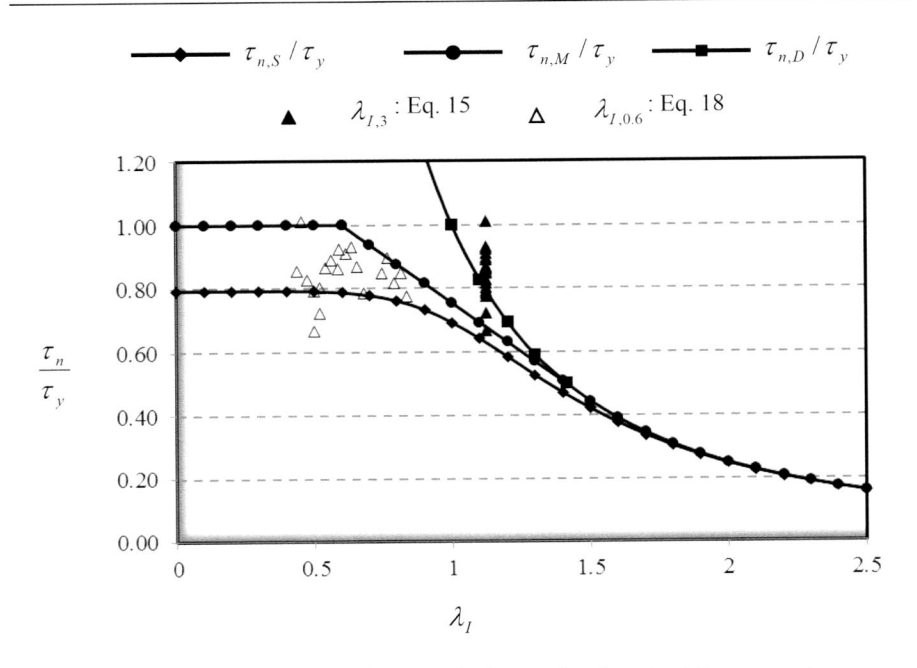

Figure 4.3 Normalized shear strength versus the interactive shear buckling parameter.

4.4 Normal-strength steel tapered girders

In a tapered bridge girder as that forming Chuhe River Bridge, shown in Fig. 2.10, various typologies of girder panels can be found. It depends on the inclination of the flange and whether the flange is under tension or compression as well as the direction of the developed tension field, which may appear on either the short or long web diagonals, as shown by Bedynek et al. [18]. Currently, these typologies are illustrated in Fig. 4.4, depending on the direction of the tension field (i.e., in long or in short diagonal) and the type of the axial force (tension or compression) of the inclined flange (Fig. 4.5). On the other hand, the current design code for plated structural elements EN 1993-1-5 [19] proposes to determine the ultimate shear resistance of tapered plate girders with flat web plates as prismatic ones. This was carefully checked recently by Bedynek et al. [18] on different types of tapered plate girders with conventional flat webs. They found that the ultimate shear resistance calculated according to EN 1993-1-5 [19] is on the safe side only for Cases I and II of flat webs, because these typologies behave similarly for equivalent rectangular plates. On the contrary, EN 1993-1-5 [19] rules were found to overestimate the ultimate shear resistance for Cases III and IV.

In this part, different typologies of linearly tapered girders with steel corrugated webs (Fig. 4.4) are investigated. In the authors' view, linearly tapered girders should be investigated first to obtain the fundamental behavior of such girders before the investigation can be extended to tapered parabolic alignment (Fig. 2.10). This is made by carrying out elastic bifurcation buckling analyses using ABAQUS software

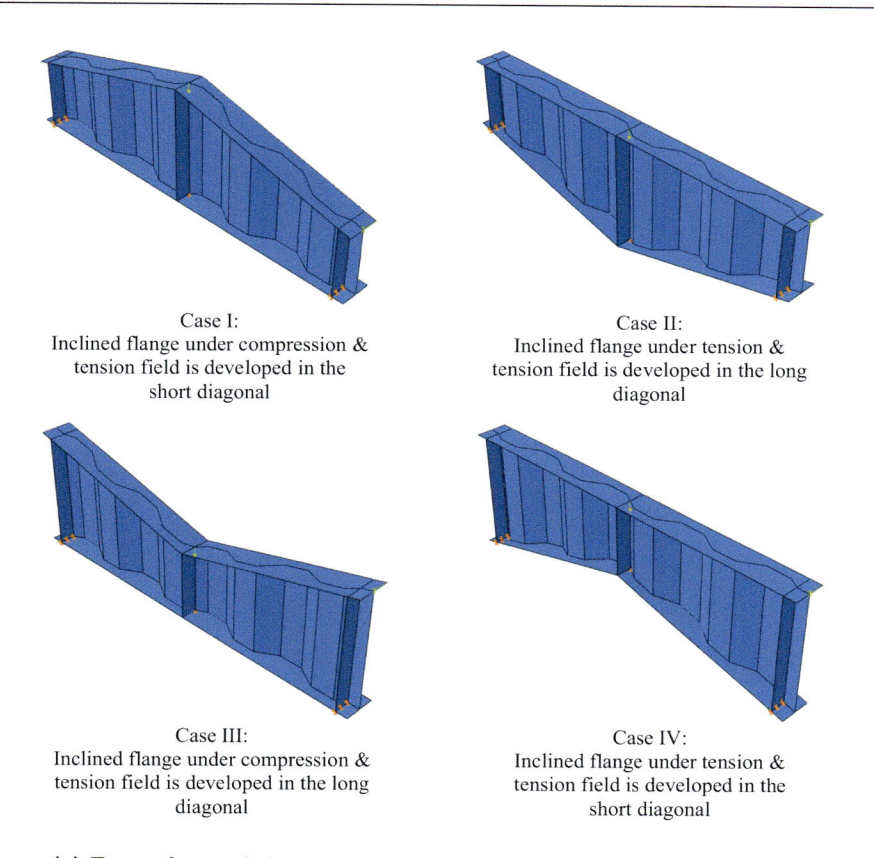

Case I:
Inclined flange under compression &
tension field is developed in the
short diagonal

Case II:
Inclined flange under tension &
tension field is developed in the long
diagonal

Case III:
Inclined flange under compression &
tension field is developed in the long
diagonal

Case IV:
Inclined flange under tension &
tension field is developed in the
short diagonal

Figure 4.4 Types of tapered plate girders with corrugated webs.

[20] on isolated corrugated webs with different boundary conditions. The considered corrugated webs have the real dimensions of two bridges; Shinkai and Matsnoki bridges. The shear strength of such *tapered* BGCWs id then examined. It is worth pointing out that the current FE models for tapered corrugated webs as well as tapered BGCWs were based on the verifications previously made by Hassanein and Kharoob [9,17].

4.4.1 Elastic bifurcation buckling analysis

4.4.1.1 General

The elastic buckling stress was calculated for the first positive Eigenmode corresponding to the shear buckling mode. S8R5 reduced integration thin shell elements were used in the current bifurcation buckling analysis. Two boundary condition cases were considered. In the first one, the four edges of the isolated corrugated plates were considered simple, as can be seen in Table 4.5. In the second case, the boundary

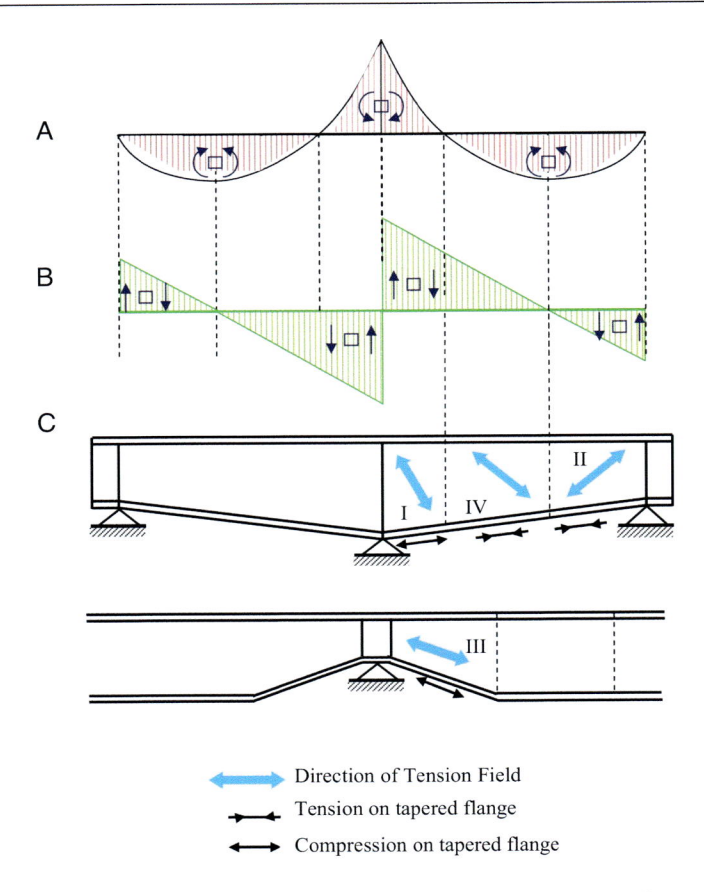

Figure 4.5 Tapered plates in continuous steel bridges; (A) bending moment diagram, (B) shear force diagram, and (C) web typology.

Table 4.5 Boundary conditions used for the isolated corrugated web plates.

Deformation	Symbols	Simple isolated plate				Fixed isolated plate			
		AB	BC	CD	DA	AB	BC	CD	DA
Translation	δ_x	R	R	R	R	R	R	R	R
	δ_y	F	F	F	R	F	F	F	R
	δ_z	R	R	R	R	R	R	R	R
Rotation	θ_x	F	F	F	F	R	F	R	F
	θ_y	F	F	F	F	R	F	R	F
	θ_z	F	F	F	F	R	F	R	F

conditions between the web and flanges of the corrugated webs were taken fixed; see Table 4.5. In the table, R stands for a restrained boundary condition, while F represents a free boundary condition. Each web plate was subjected to an even vertical load at edge BC, as can be seen in Fig. 4.6, which ensures pure shear loading. It is worth pointing

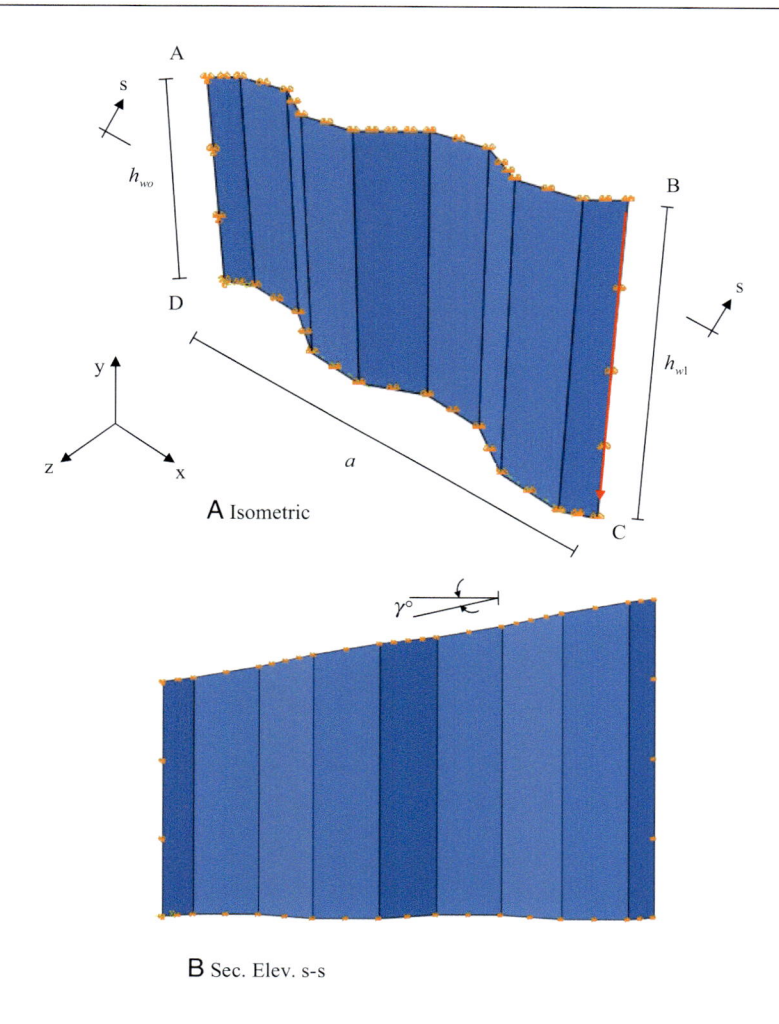

A Isometric

B Sec. Elev. s-s

Figure 4.6 Boundary conditions and load application of a typical corrugated web: (A) isometric and (B) Sec. Elev. s-s.

out that h_{w1} and h_{wo} represent the long and short vertical ends of the corrugated web of any typology (or the shear span of the girder), respectively. Young's modulus of 210 GPa and a Poisson's ratio of 0.3 were used all through the current analyses. Additional information could be found in [9].

4.4.1.2 Input data

A parametric study is presented herein. Fig. 4.1, which presents the typical corrugation configuration, represents the case of tapered girders. Regarding the geometry, the corrugation dimensions of considered corrugated webs were taken typical of those used in Shinkai and Matsnoki bridges, as given in Table 3.1. Taking only two corrugation dimensions was found to be sufficient following the results of Nie et al. [21]. In their study, they found that the relationship between the normalized critical stress and the slenderness parameter *is unrelated* to the corrugated configuration and shear span ratio. However, based on previous research, the trapezoidally corrugated steel web plates need to satisfy the geometric conditions of $b/h_w \leq 0.2$ [2,5] and $h_r/t_w \geq 10$ [21]. Therefore, the web heights of control prismatic corrugated webs were taken as 1250 and 1500 mm for current Shinkai and Matsnoki webs, respectively, instead of the values shown in Table 3.1. By doing so, the limit value of 0.2 for b/h_w was considered. The web thickness (t_w) was chosen to have an h_r/t_w ratio greater than and include 10.7, as can be seen later. Each plate was assumed to consist of two corrugation waves ($a = 4b + 4d$). This results in a ratio of a/h_{w1} greater than unity, which is in accordance with the conditions of corrugated web girders used in practice, where the distance between the vertical stiffeners is much greater than the web depth (h_w). The current study contains 260 isolated webs, as shown in Table 4.5, covering five main parameters: corrugation configuration, typology of the tapered web plate, web thickness (t_w), inclination angle of the upper or lower edges (γ), and the boundary condition of the upper or lower edges (simple or fixed).

4.4.1.3   Buckling modes

Before FE program described above was generated, different web thicknesses were used to obtain the three different modes that may be observed in the corrugated webs. The difference between this preliminary analysis and FE program explained in Section 4.3.2 was the web thickness values. In the preliminary analysis, a wide range of t_w values was considered starting from very small web thickness to a very high value giving $h_r/t_w < 10$ which is not applicable to bridges. Fig. 4.7 provides the local, interactive, and global buckling modes for both cases of prismatic and tapered corrugated webs of Case I using the corrugation configuration of Shinkai bridge. The other cases of tapered plates (shown in Fig. 4.4) were also generated and they provided qualitatively the same buckling modes, so they were not added to Fig. 4.7 for brevity. As can be seen, the local buckling mode is controlled by deformations within a single subpanel (fold) of the web and it was found to occur for very slender webs. On the opposite, the global buckling mode involves multiple folds and the buckled shape extends diagonally over the depth of the web. This global buckling mode was found to occur for plates with high t_w values. However, the interactive buckling mode appears to have characteristics of both local and global buckling modes. Compared to prismatic webs, the tapered webs provided the same buckling patterns (Fig. 4.7). However, the local and interactive buckling waves of the tapered plates were practically shifted toward the short vertical ends of the plates.

In the current control models which have the corrugation dimensions of real bridges, it was found that only interactive and global buckling modes occur in prismatic webs. For both corrugations, the interactive buckling mode was observed for web thicknesses of 6 and 8 mm, whereas the buckling was global for higher web thicknesses (10, 12, and 14 mm). On the other hand, the entire tapered corrugated webs were buckled

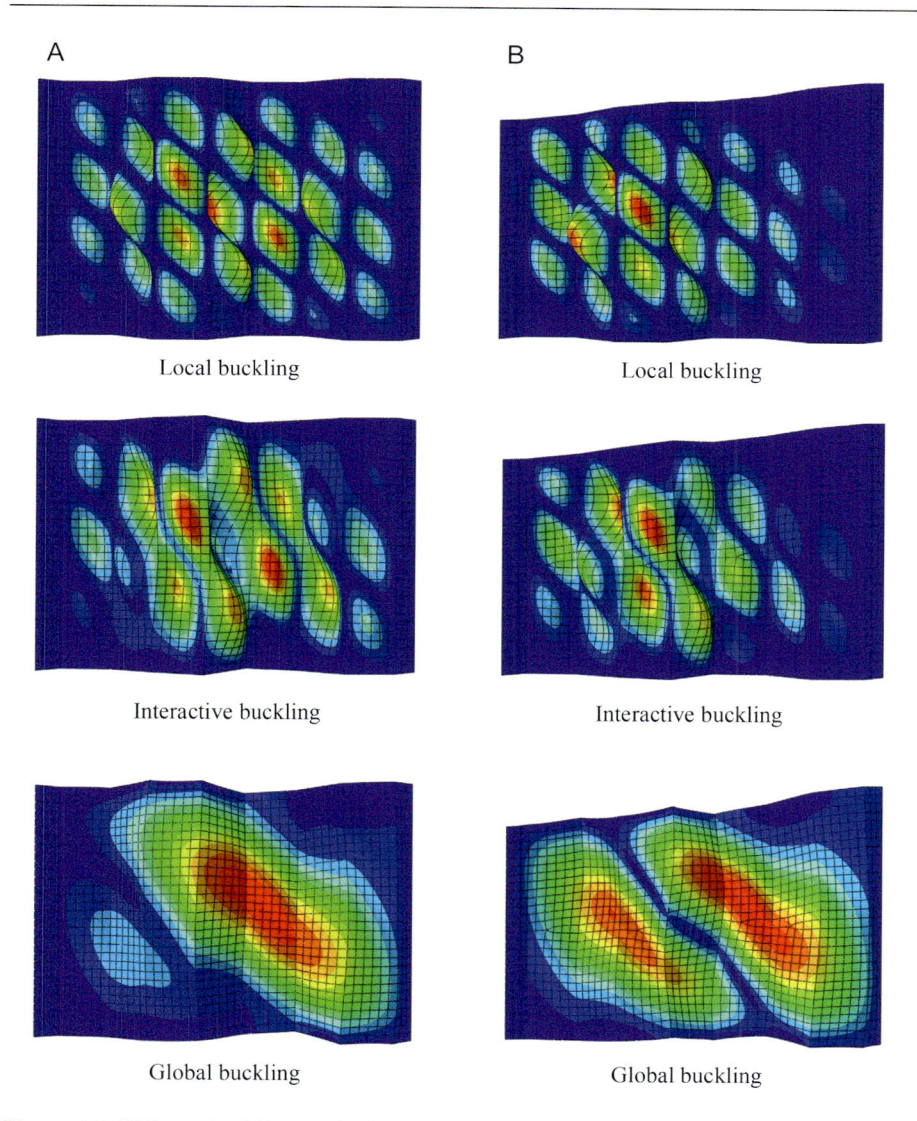

A

B

Local buckling   Local buckling

Interactive buckling   Interactive buckling

Global buckling   Global buckling

Figure 4.7 Different buckling modes in (A) prismatic and (B) tapered corrugated webs.

interactively, with five exceptions related to the inclination angle of $15°$ and $t_w = 6$ mm which buckled locally. This means that the global buckling disappears in tapered girders (with different typologies) from small inclination angles. Fig. 4.8 provides the interactive buckling modes of different typologies. As can be seen, the buckled waves are shifted toward the short vertical edge in the different cases. However, the buckled waves are slightly stretched when the buckling is in the direction of the long diagonal of the corrugated plate (Cases II and III) compared to the other cases.

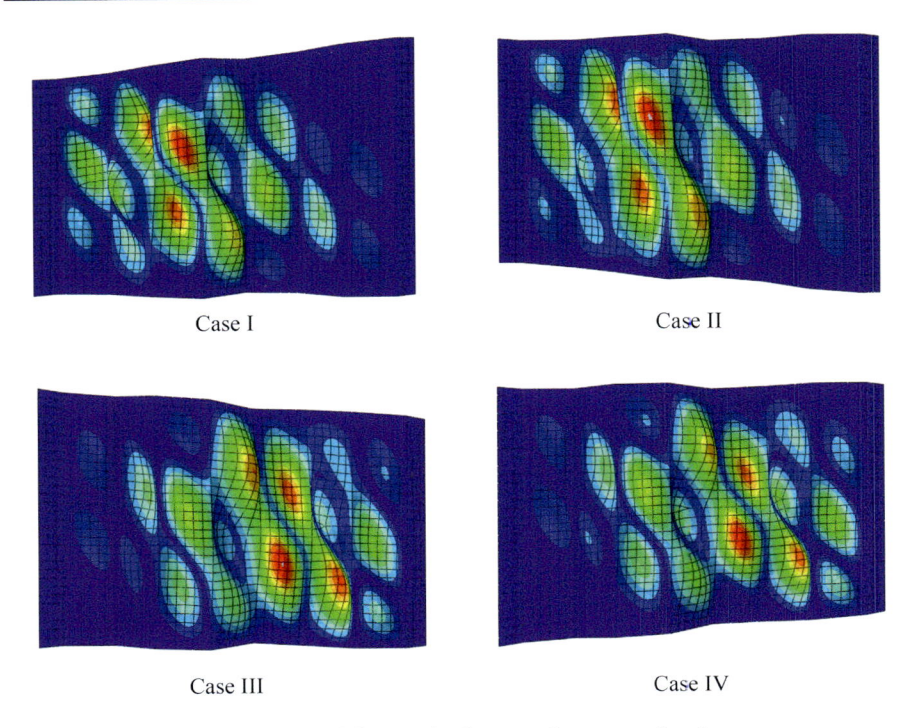

Case I   Case II

Case III   Case IV

Figure 4.8 Typical interactive buckling modes in tapered corrugated webs.

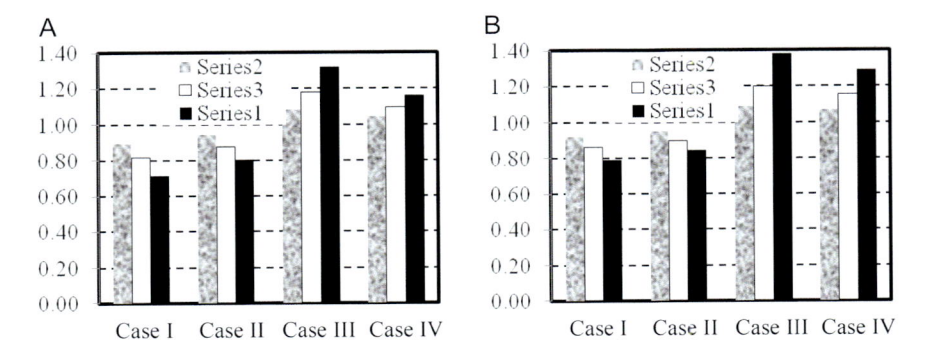

Figure 4.9 Normalized critical stresses for tapered corrugated webs: (A) $t_w = 6$ mm and (B) $t_w = 14$ mm.

4.4.1.4 *Effects of tapered web typology*

As seen above, the buckling patterns of different tapered corrugated webs, having the same corrugation dimensions, are the same. On the other hand, the values of the critical shear stress ($\tau_{cr,FE}$) were found to differ between them. Fig. 4.9 illustrates the normalized critical stresses for different typologies calculated with respect to the

Table 4.6 Proposed to finite element critical buckling stresses ($\tau_{cr,\mathrm{Prop}}/\tau_{cr,FE}$) ratios.

t_w y_o [mm]	Case I		Case II		Case III		Case IV	
	Shinkai	Matsnoki	Shinkai	Matsnoki	Shinkai	Matsnoki	Shinkai	Matsnoki
5 6	1.03	1.01	1.01	1.02	1.01	1.01	0.99	0.98
10 6	1.04	1.01	1.01	1.01	1.03	1.02	1.04	1.03
15 6	1.10	1.02	1.02	1.03	1.04	1.02	1.10	1.09
5 8	1.01	1.01	1.01	1.01	1.01	1.01	0.97	0.96
10 8	1.00	0.99	0.99	0.99	1.01	1.00	1.00	0.99
15 8	1.04	0.99	0.99	0.99	1.00	0.99	1.04	1.02
5 10	1.00	1.01	1.01	1.01	1.01	1.01	0.97	0.96
10 10	0.99	0.99	0.99	0.99	1.01	1.00	0.99	0.98
15 10	1.01	0.98	0.98	0.98	0.99	0.98	1.00	0.99
5 12	1.00	1.01	1.01	1.01	1.01	1.01	0.97	0.96
10 12	0.99	0.99	0.99	0.99	1.01	1.00	0.99	0.98
15 12	1.00	0.97	0.97	0.98	0.99	0.98	1.00	0.98
5 14	1.00	1.01	1.01	1.01	1.01	1.01	0.97	0.96
10 14	0.99	0.99	0.99	0.99	1.01	1.00	0.99	0.98
15 14	1.00	0.97	0.97	0.98	0.99	0.98	1.00	0.98
Ave	1.01	1.01	1.00	1.00	1.01	1.01	1.00	0.99
COV	0.030	0.030	0.016	0.015	0.013	0.013	0.037	0.034

prismatic web with the same web thickness. Plates with the extreme t_w values were presented in the figure (6 and 14 mm) with the Shinkai corrugation dimensions. It can be seen from the figure that the $\tau_{cr,FE}$ value becomes less than the value of the prismatic web for Cases I and II, while it is higher for Cases III and IV. Hence, it could be concluded that the predictions of the τ_{cr} value for tapered webs based on prismatic web calculations may overestimate the stresses for Cases I and II, whereas it may provide highly conservative values for Cases III and IV. It can additionally be seen from the figure that by increasing the web thickness, the normalized critical stresses for the different typologies increase.

By analyzing the $\tau_{cr,FE}$ values of the corrugated webs, it was possible to correlate them with the critical values of the prismatic webs ($\tau_{cr,FE,P}$). Accordingly, the following equations are proposed to predict the critical buckling stresses:

$$\tau_{cr,\mathrm{Prop}} = \tau_{cr,FE,P}/(1 + \tan\gamma)\ \text{for Case I} \tag{4.15}$$

$$\tau_{cr,\mathrm{Prop}} = 1.04\tau_{cr,FE,P}/(1 + \tan\gamma)\ \text{for Case II} \tag{4.16}$$

$$\tau_{cr,\mathrm{Prop}} = \tau_{cr,FE,P}/(1 - \tan\gamma)\ \text{for Case III} \tag{4.17}$$

$$\tau_{cr,\mathrm{Prop}} = 0.94\tau_{cr,FE,P}/(1 - \tan\gamma)\ \text{for Case IV} \tag{4.18}$$

It can be observed that the proposed critical shear stresses for Cases II and IV were additionally adjusted to accord with FE results. Table 4.6 provides the ratios of the proposed critical shear stress ($\tau_{cr,\mathrm{Prop}}$) to FE critical shear stress of the tapered webs ($\tau_{cr,\mathrm{Prop}}/\tau_{cr,FE}$). As can be seen, the above equations can effectively be used to predict

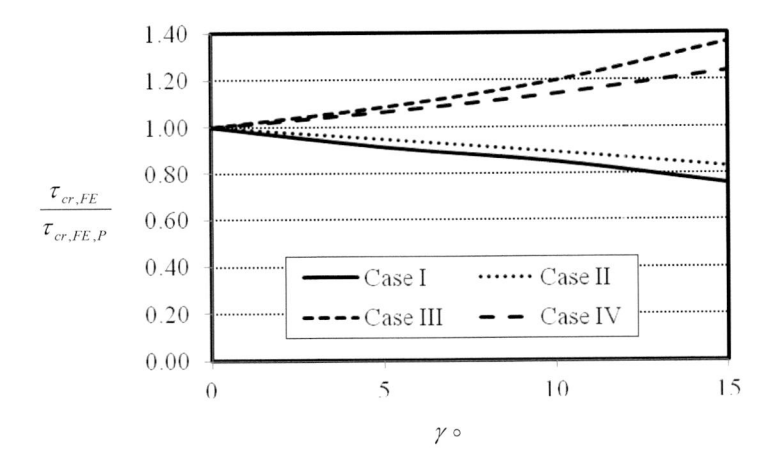

Figure 4.10 Normalized critical stresses vs. angle of inclination with Shinkai corrugation configuration.

the critical buckling stresses of tapered corrugated webs provided that the critical values of the prismatic webs ($\tau_{cr,FE,\,P}$) are known.

4.4.1.5 Effects of inclination angle of the upper or lower edges

The angle of inclination (γ°) in different typologies was found to affect their critical shear stresses ($\tau_{cr,FE}$). Fig. 4.10 shows the relationship between the critical shear stress of tapered webs relative to those of the prismatic webs ($\tau_{cr,FE}/\tau_{cr,FE,\,P}$) and the angle of inclination (γ°). The webs having $t_w = 8$ mm and the dimensions of Shinkai corrugation were considered. As can be observed, increasing the value of γ° reduces the $\tau_{cr,FE}$ values for Cases I and II. On the opposite, the values of $\tau_{cr,FE}$ increase by increasing the value of γ° for Cases III and IV.

4.4.1.6 Effects of web thickness

Increasing the thickness of the corrugated webs in the four typologies raises the critical shear stress ($\tau_{cr,FE}$); see Fig. 4.11. This figure shows the relationship between $\tau_{cr,FE}$ and the web thickness values for different typologies. As can be seen, the increase in $\tau_{cr,FE}$ is almost linear by increasing the web thickness of the tapered corrugated webs.

4.4.1.7 Effects of web thickness

4.4.1.7.1 Prismatic corrugated webs
In this section, the critical shear values of the prismatic webs ($\tau_{cr,FE,\,P}$) are compared with the available critical stresses proposed by Yi. et al. [22] which was introduced in Eq. (3.8) with $n = 1.0$; $\tau_{cr,l,\,1}$. Two values for the global shear buckling coefficient (k_G) of 36 [21] and 31.6 [2] were used for the case of simple juncture. Additionally, the second-order interactive buckling strength ($\tau_{cr,l,\,2}$) according to Abbas et al. [23] was

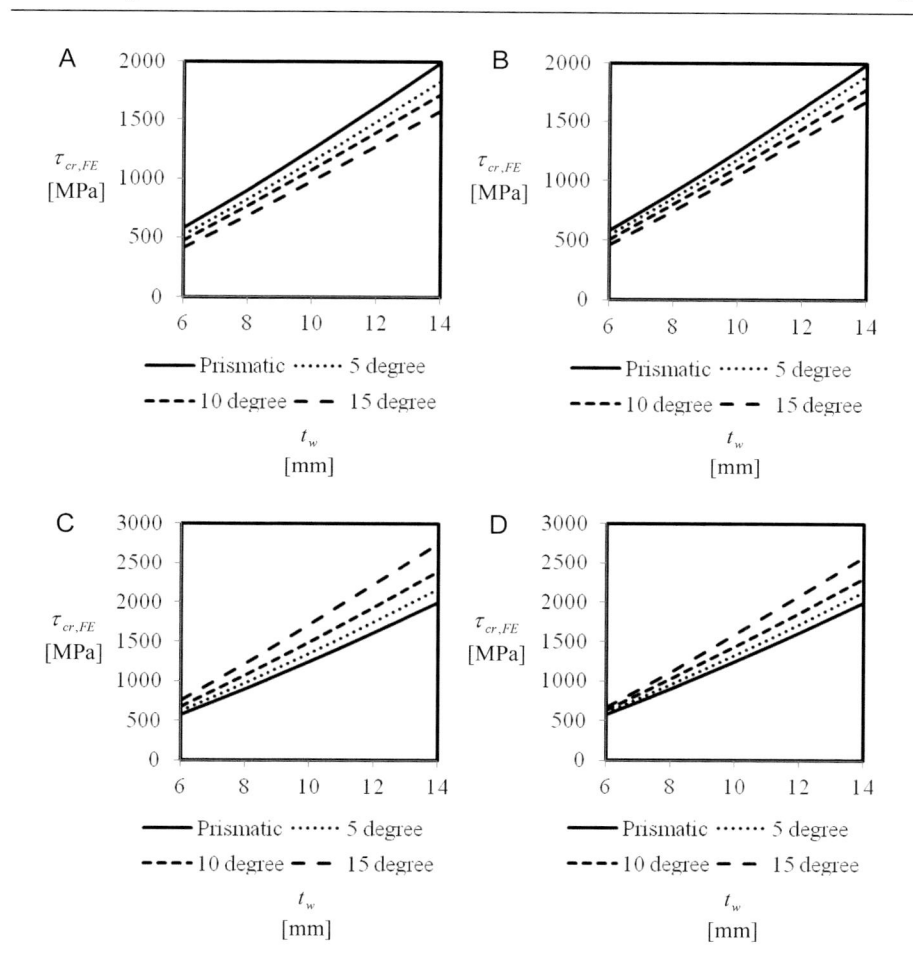

Figure 4.11 FE critical stresses for tapered corrugated webs with Shinkai corrugation configuration: (A) Case I, (B) Case II, (C) Case III, and (D) Case IV. *FE*, finite element.

calculated (using Eq. (3.8) with $n = 2.0$). The results of the comparison are provided in Table 4.7. From the comparative results, it can be noticed that the first-order interactive buckling strength ($\tau_{cr,I,\,1}$) proposed by Yi. et al. [22] is *suitable* for the case of corrugated webs with both values of k_G. With $k_G = 36$, the buckling stresses of the prismatic corrugated webs are well calculated compared to the other value. On the contrary, $\tau_{cr,I,\,2}$ provides unsafe results.

4.4.1.7.2 Tapered corrugated webs
Here, the critical shear values of the tapered webs ($\tau_{cr,FE}$) are compared to the proposed Eqs. (4.15–4.18) using $\tau_{cr,FE,\,P} = \tau_{cr,I,\,1}$. As can be seen from Table 4.8, the current proposed equations, which utilize the first-order interactive buckling strength ($\tau_{cr,I,\,1}$) proposed by Yi. et al. [22] are suitable for use in the prediction equations of the shear

Table 4.7 Comparison between the critical shear stress ($\tau_{cr,FE,\,P}$) of the prismatic corrugated webs with first- and second-order interactive buckling strength (simple juncture).

h_w	t_w	$\tau_{cr,FE,\,P}/\tau_{cr,I,\,1}$		$\tau_{cr,FE,\,P}/\tau_{cr,I,\,2}$	
[mm]	[mm]	$k_G = 36$	$k_G = 31.6$	$k_G = 36$	$k_G = 31.6$
1250	6	1.17	1.19	1.00	1.01
1250	8	1.11	1.14	0.90	0.91
1250	10	1.07	1.11	0.83	0.84
1250	12	1.05	1.10	0.78	0.80
1250	14	1.03	1.09	0.75	0.78
1500	6	1.16	1.18	0.99	1.00
1500	8	1.10	1.13	0.89	0.90
1500	10	1.06	1.10	0.81	0.83
1500	12	1.04	1.09	0.77	0.79
1500	14	1.02	1.08	0.74	0.77
Ave		1.08	1.12	0.85	0.86
COV		0.052	0.040	0.097	0.087

Table 4.8 Proposed to finite element critical buckling stresses ($\tau_{cr,\mathrm{Pr}\,op}/\tau_{cr,FE}$) ratios.

γ_0	t_w [mm]	Case I		Case II		Case III		Case IV	
		Shinkai	Matsnoki	Shinkai	Matsnoki	Shinkai	Matsnoki	Shinkai	Matsnoki
5	6	0.88	0.87	0.87	0.88	0.87	0.87	0.85	0.85
10	6	0.89	0.90	0.86	0.87	0.88	0.88	0.89	0.89
15	6	0.95	0.96	0.88	0.89	0.89	0.88	0.95	0.94
5	8	0.91	0.90	0.91	0.92	0.91	0.92	0.87	0.88
10	8	0.90	0.91	0.89	0.90	0.91	0.91	0.90	0.90
15	8	0.94	0.95	0.89	0.90	0.90	0.90	0.93	0.93
5	10	0.94	0.93	0.94	0.96	0.94	0.95	0.90	0.91
10	10	0.92	0.94	0.92	0.94	0.94	0.95	0.93	0.93
15	10	0.94	0.96	0.91	0.93	0.93	0.92	0.94	0.93
5	12	0.96	0.96	0.97	0.98	0.97	0.97	0.92	0.93
10	12	0.94	0.96	0.94	0.96	0.97	0.97	0.95	0.95
15	12	0.96	0.97	0.93	0.94	0.95	0.94	0.95	0.95
5	14	0.97	0.97	0.98	0.99	0.98	0.98	0.94	0.94
10	14	0.96	0.97	0.96	0.97	0.98	0.98	0.96	0.96
15	14	0.97	0.98	0.94	0.96	0.96	0.96	0.96	0.96
Ave		0.93	0.94	0.92	0.93	0.93	0.93	0.92	0.92
COV		0.028	0.032	0.036	0.037	0.036	0.038	0.034	0.032

buckling stresses of the tapered corrugated webs. The $\tau_{cr,\mathrm{Pr}\,op}$ values, in average, are about 94% of FE critical stresses ($\tau_{cr,FE}$).

4.4.1.7.3 Effect of fixed web-to-flange junctures

Since the boundary condition between the web and flanges of *prismatic* corrugated webs was found to effectively influence the behavior of the girders [9], the upper boundary condition limit (i.e., fixed) was also taken into account in this extension.

Table 4.9 Proposed to finite element critical buckling stresses ($\tau_{cr,\mathrm{Prop}}/\tau_{cr,FE}$) ratios for corrugated webs with fixed juncture with flanges.

γ_0	t_w [mm]	Case I		Case II		Case III		Case IV	
		Shinkai	Matsnoki	Shinkai	Matsnoki	Shinkai	Matsnoki	Shinkai	Matsnoki
5	6	0.88	0.86	0.86	0.87	0.86	0.86	0.84	0.84
10	6	0.88	0.89	0.85	0.86	0.87	0.87	0.88	0.88
15	6	0.93	0.94	0.86	0.87	0.87	0.86	0.93	0.93
5	8	0.88	0.87	0.88	0.89	0.88	0.88	0.85	0.85
10	8	0.88	0.89	0.86	0.87	0.88	0.88	0.88	0.88
15	8	0.91	0.92	0.86	0.87	0.87	0.87	0.91	0.91
5	10	0.90	0.89	0.90	0.91	0.90	0.91	0.87	0.87
10	10	0.89	0.90	0.88	0.89	0.90	0.90	0.89	0.89
15	10	0.91	0.92	0.87	0.88	0.89	0.88	0.90	0.90
5	12	0.92	0.91	0.92	0.93	0.92	0.93	0.88	0.89
10	12	0.90	0.91	0.90	0.91	0.92	0.92	0.90	0.90
15	12	0.91	0.92	0.88	0.89	0.90	0.89	0.91	0.90
5	14	0.93	0.92	0.93	0.94	0.93	0.94	0.89	0.90
10	14	0.91	0.92	0.91	0.92	0.93	0.93	0.91	0.91
15	14	0.92	0.93	0.89	0.90	0.91	0.90	0.92	0.91
Ave		0.90	0.91	0.88	0.89	0.90	0.89	0.89	0.89
COV		0.019	0.023	0.024	0.025	0.023	0.025	0.026	0.024

Table 4.1 Provides the boundary conditions used in the models. The critical shear buckling stresses were then compared with the predictions suggested above for the case of tapered corrugated webs with simple boundary conditions. The comparative results are given herein in Table 4.9. As can be seen, the proposed equations may be considered as suitable predictors for tapered corrugated webs with a fixed upper and lower junctures. The average $\tau_{cr,\mathrm{Prop}}/\tau_{cr,FE}$ ratios are 0.90 for different typologies. As the web junctures with the flanges, in reality, lie between simple and fixed conditions, the proposed equations can be used in practice for the case of tapered corrugated girders (with flanges) without the need for new equations as made by Hassanein and Kharoob [9].

4.4.2 Nonlinear buckling analysis

4.4.2.1 General

A two-step approach was used in the nonlinear simulation of BGCWs to include the initial geometric imperfections. In the first step, an elastic buckling analysis was performed on a perfect BGCW to obtain the buckling mode. In the second step, the initial geometric imperfections based on the first buckling mode were included in the nonlinear analysis of BGCW under mid-span concentrated load using the modified RIKS method. S8R5 reduced integration thin shell elements were employed to discretize the models in the current nonlinear analysis. Simply supported boundary conditions were applied to end sections. The steel material has been modeled as a von Mises material with isotropic hardening. The steel used was S355 according to EN 1993-1-1 [24], which has a yield (f_y) and an ultimate strength (f_u) of 355MPa and 510MPa, respectively. A bilinear elastic–plastic stress–strain curve with linear strain hardening was used to simulate the steel material. Additional information could be found in [17] for the same authors.

4.4.2.2 Input data

Numerical program was conducted here on full-scale BGCWs subjected to shear loading to substitute the lack of their shear behavior and strength. Each girder consisted of two flat compact flanges and a corrugated web. As used in [17], the width and thickness of the flat flanges were chosen as 500 mm and 50 mm, respectively, ensuring that the $t_f/t_w \geq 3.0$. The Shinkai corrugation dimensions (Table 4.2) were used herein. The web depth of the prismatic girders was fixed to 1250 mm which is equal to dimension (h_{w1}) of tapered girders. The shear span (a) consisted of four corrugation waves ($a = 8b + 8d = 3600$ mm). Hence, the aspect ratio of the web panel (a/h_w) was greater than unity, which is consistent with the conditions of practical BGCWs, where the distance between the vertical stiffeners is much greater than the web depth (h_w). The shear buckling parameter of the corrugated webs ($\lambda_s = \sqrt{\tau_y/\tau_l}$) was calculated; τ_l was calculated using Eqs. (4.15–4.18) for the different types. As can be seen in Table 4.8, λ_s is lower than 0.6 for the considered BGCWs except for those with $t_w = 6$ mm. The four tapered typologies were taken into consideration with $h_{wo} = 800$ mm. This resulted in an inclination angle of 7.125° for all typologies. The web thickness varied from 6 to 14 mm with an increment of 2 mm. This resulted in five prismatic and twenty tapered BGCWs.

4.4.2.3 Results and discussions

The results of the current nonlinear buckling analyses are the failure modes, the load versus mid-span deflection relationships and ultimate shear strengths ($V_{ul,FE}$). Table 4.10 provides FE results. The maximum shear stress values ($\tau_{ul,FE}$) at the critical cross-section of BGCWs were calculated by Eq. (4.1) using the shorter web dimension (h_{wo}) for all cases. The relative ratios of $\tau_{ul,FE}/\tau_y$ were also calculated; where τ_y is the yield shear strength of the base material ($\tau_y = f_y/\sqrt{3}$).

The full-scale BGCWs were loaded here to fail under shear except for the girder of Case II with $t_w = 14$ mm which failed by the flange yielding (flexural failure (F)). The failure modes developed were the interactive (I) and the global (G) shear buckling. As can be seen in Table 4.8, interactive shear buckling was associated with BGCWs with small values of t_w, while the global mode appeared in girders with relatively high values of t_w. It could additionally be observed that the failure mode of the tapered BGCWs with $t_w = 10$ mm was changed to interactive instead of the global mode which was found in the prismatic girders. This means that making BGCW tapered may change the failure mode compared to the prismatic girder. Fig. 4.12 shows the stress contour at the ultimate shear loads for some selected girders. The gray color represents the portions

Table 4.10 Finite element results.

Type	t_w [mm]	λ_s	$V_{ul,FE}$ [kN]	Buckling mode	$\tau_{ul,FE}$ [MPa]	$\frac{\tau_{ul,FE}}{\tau_y}$	$\frac{\tau_{ul,M}}{\tau_y}$	$\frac{\tau_{ul,S}}{\tau_y}$
Prismatic	6	0.644	1247	I	166	0.81	0.95	0.78
	8	0.503	1783	I	178	0.87	1.00	0.79
	10	0.419	2258	I	181	0.88	1.00	0.79
	12	0.364	2737	G	182	0.89	1.00	0.79
	14	0.326	3325	G	190	0.93	1.00	0.79
Case I	6	0.683	944	I	197	0.96	0.94	0.78
	8	0.533	1318	G	206	1.00	1.00	0.79
	10	0.445	1693	G	212	1.03	1.00	0.79
	12	0.386	2144	G	223	1.09	1.00	0.79
	14	0.346	2619	G	234	1.14	1.00	0.79
Case II	6	0.696	1097	I	229	1.12	1.00	0.79
	8	0.544	1619	G	253	1.23	1.00	0.79
	10	0.453	2010	G	251	1.23	1.00	0.79
	12	0.394	2147	G	224	1.09	1.00	0.79
	14	0.352	2295	F	205	1.00	1.00	0.79
Case III	6	0.602	652	I	136	0.66	0.99	0.79
	8	0.470	917	I	143	0.70	1.00	0.79
	10	0.392	1146	G	143	0.70	1.00	0.79
	12	0.341	1450	G	151	0.74	1.00	0.79
	14	0.305	1686	G	150	0.73	1.00	0.79
Case IV	6	0.624	634	I	132	0.64	0.95	0.78
	8	0.487	894	I	140	0.68	1.00	0.79
	10	0.406	1136	G	142	0.69	1.00	0.79
	12	0.353	1397	G	145	0.71	1.00	0.79
	14	0.316	1666	G	149	0.73	1.00	0.79

exceeding the yield strength of the material. These portions are concentrated in the side including the short depth of the webs (h_{wo}) where the shear stress is maximum. Nearly one corrugation wave, on each shear span (a), was yielded in different cases.

Regarding the ultimate shear strengths ($V_{ul,FE}$) of BGCWs, it can be seen from Table 4.8 that tapered girders have always reduced $V_{ul,FE}$ values compared to prismatic ones. However, the highest $V_{ul,FE}$ value was found in Case II (case of long diagonal & inclined flange under tension). Case I has a lower value compared to Case II, while Cases III and IV have the least $V_{ul,FE}$ values among the different types. The results also show that BGCWs with Cases III and IV have nearly the same shear strengths. This means that neither the type of the axial force in the inclined flange (tension or compression) nor the direction of the developed tension field (short or on the long web diagonal) has any effect on the strength of these girders (Cases III and IV). On the other hand, apparently Cases I and II can reach $\tau_{ul,FE}/\tau_y$ ratios greater than unity; see Fig. 4.13. This means that their materials can effectively be utilized, which is not the case of BGCWs with Cases III and IV. However, the increase of the t_w value increases the strengths of BGCWs linearly except for Case II; see Fig. 4.14.

A   Case I

B   Case II

C   Case III

D   Case IV

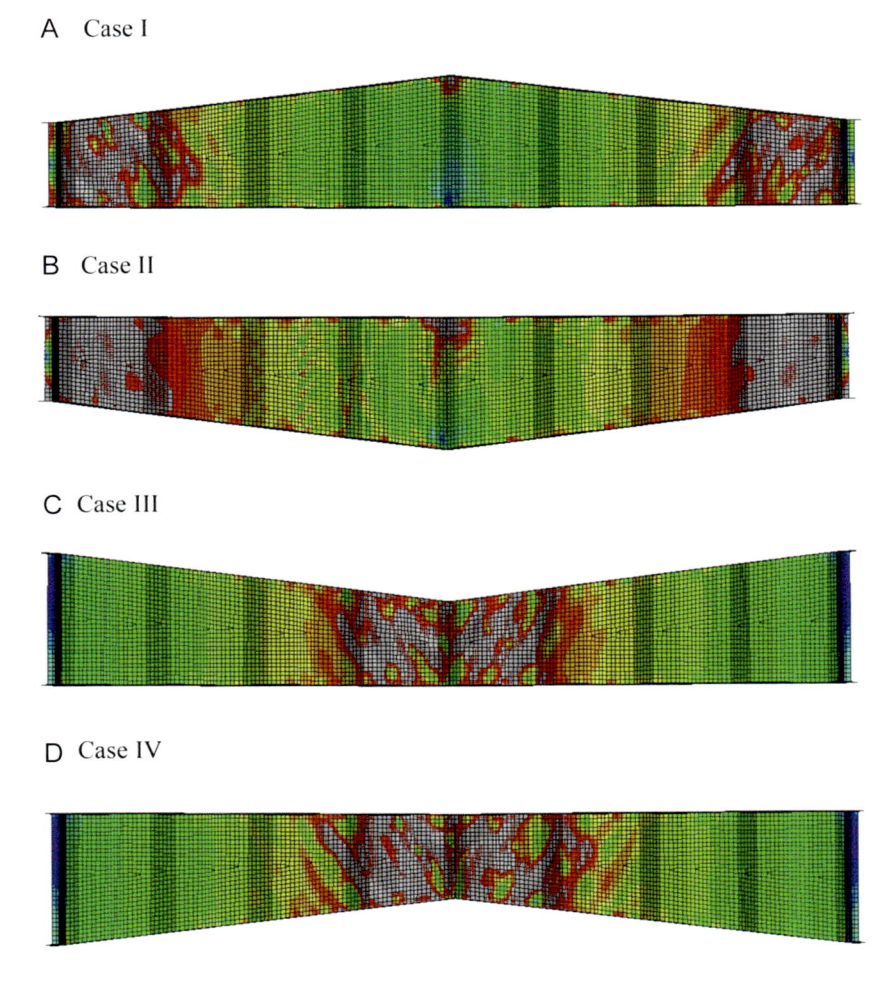

Figure 4.12 Stress contour for different BGCWs with 10 mm web thickness. *BGCWs*, bridge girders with corrugated webs.

A sample of the load-mid-span deflection for BGCWs can be seen in Fig. 4.15. Case IV as a sample was considered here because it was found that others have qualitatively similar relationships. It can be observed that the failure is sudden and results from the buckling of BGCW with small web thickness (i.e. $t_w = 6$ mm). Tapered BGCWs with higher t_w values exhibit a smooth transition from the pre-peak to the post-peak stages. It is worth pointing out that this is different from the behavior of prismatic girders found by Hassanein and Kharoob [17] where the failure was found to occur suddenly for different web thicknesses. The load-carrying capacity decreases after the maximum load is achieved, but some residual strength remains after failure. In addition,

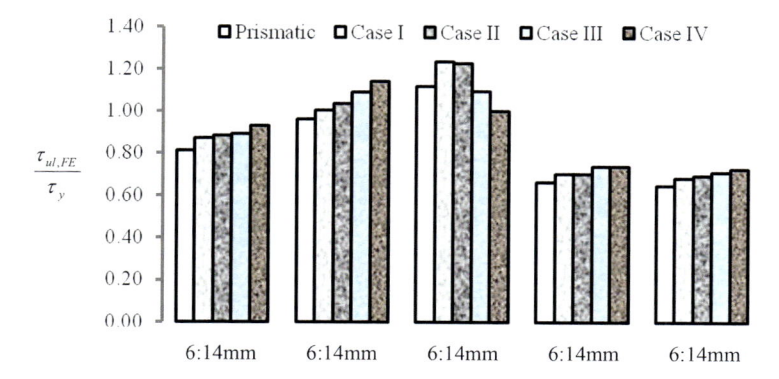

Figure 4.13 Comparison between $\tau_{ul,FE}/\tau_y$ ratios for different groups.

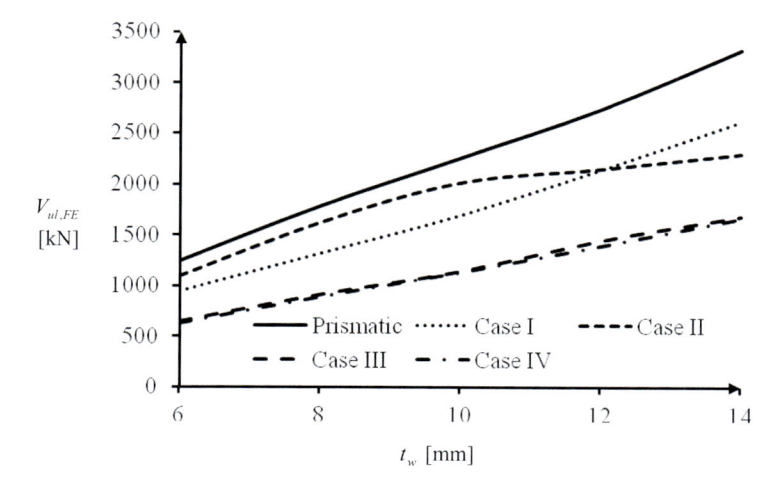

Figure 4.14 Shear ultimate strengths vs. the web thickness for different groups.

an obvious decrease in the initial stiffness of the girder is associated with the decrease of its corrugated web thickness.

4.4.2.4 *Comparisons with available shear strengths*

The ultimate strengths of the current models were compared with the shear strengths proposed by Moon et al. [10] (Eq. (4.19)) and Sause and Braxtan [11] (Eq. (4.20)) for prismatic BGCWs. This was made using the current shear buckling parameter of corrugated webs ($\lambda_s = \sqrt{\tau_y/\tau_I}$) instead of their parameters. The results are provided in Table 4.8. As can be seen, the $\tau_{ul,M}$ values according to Moon et al. [10] provides accurate and better estimates for Cases I and II than the results of Sause and Braxtan

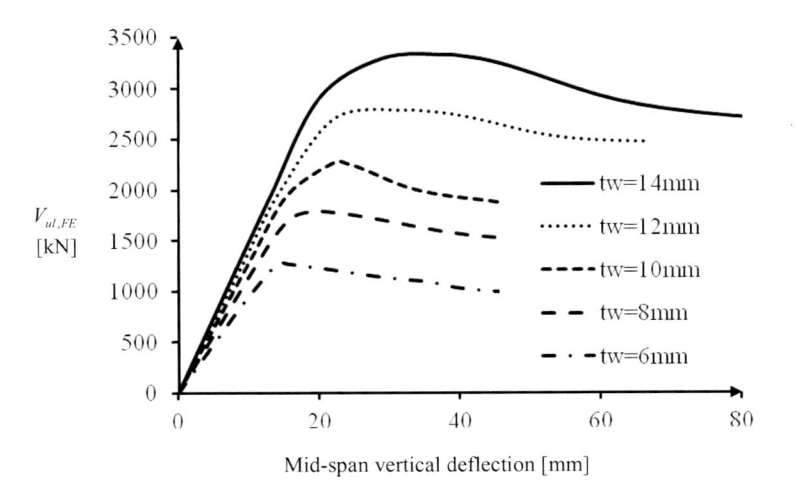

Mid-span vertical deflection [mm]

Figure 4.15 Shear ultimate strengths vs. the mid-span vertical deflection for Case IV.

[7] ($\tau_{ul,S}$). On the other hand, both strengths [7,10] overestimate the strengths of Cases III and IV. However, until a precise shear strength for Cases III and IV is proposed (based on large parametric studies for both of them), the strength by Moon et al. [10] may be used by reducing it ($\tau_{ul,M,\,\mathrm{mod}}$) by the ratio h_{wo}/h_{w1} as previously suggested by Bedynek et al. [18]; see Table 4.11.

$$\frac{\tau_{ul,M}}{\tau_y} = \begin{cases} 1.0 & : \lambda_s \leq 0.6 \\ 1 - 0.614(\lambda_s - 0.6): 0.6 < \lambda_s \leq \sqrt{2} \\ \frac{1}{\lambda_s^2} & : \sqrt{2} < \lambda_s \end{cases} \tag{4.19}$$

$$\tau_{ul,S} = \tau_y \left(\frac{1}{(\lambda_s)^6 + 2} \right)^{1/3} \tag{4.20}$$

Table 4.11 Suggested strengths for tapered bridge girders with corrugated webs of Cases III and IV.

Type	t_w [mm]	λ_s	$V_{ul,FE}$ [kN]	Buckling mode	$\tau_{ul,FE}$ [MPa]	$\frac{\tau_{ul,FE}}{\tau_y}$	$\frac{\tau_{ul,M,\,\mathrm{mod}}}{\tau_y}$
Case III	6	0.602	652	I	136	0.66	0.64
	8	0.470	917	I	143	0.70	0.64
	10	0.392	1146	G	143	0.70	0.64
	12	0.341	1450	G	151	0.74	0.64
	14	0.305	1686	G	150	0.73	0.64
Case IV	6	0.624	634	I	132	0.64	0.62
	8	0.487	894	I	140	0.68	0.64
	10	0.406	1136	G	142	0.69	0.64
	12	0.353	1397	G	145	0.71	0.64
	14	0.316	1666	G	149	0.73	0.64

4.5 High-strength steel prismatic girders

High-strength steels (HSSs) are currently utilized in many structural engineering applications. These HSSs are well suited to heavily-loaded columns [25–26], girders [27–28], and beam–columns [29] because they are characterized by their relatively high strengths while their weights are minimized. Therefore, HSSs have been widely used in bridges [30]. Generally, HSSs represent those steels having a minimum proof/yield stress (F_y) of 460MPa. Another feature that makes HSSs well suited for use in top countries in terms of manufacture [31], such as China, the USA, Japan, Germany, etc., is that the consumption of the steel material and the emission of the carbon dioxide from the construction sector are reduced by utilizing HSSs [32–33]. According to World steel association (WSA) [34], using HSSs rather than NSSs reduces the steel material total release of 0.156 billion tons CO_2 equivalents. Hence, substituting NSSs by HSSs in future structural applications becomes necessary to protect the planet by reducing the carbon dioxide released from the production process of the steel. Based on above, the extensive production of HSS matches the vision of WSA [35] recently issued, aligned with the aims of the Paris Agreement [36], that the steel industry should develop advanced steel products to enable societal transition to low-carbon steel-making effectively. One of HSSs used globally nowadays is S460, which has a yield strength of 460MPa [24]. A structural example that uses this HSS grade is the bridge erected in Mttådalen, Sweden in 1995 [30]. Accordingly, this section is emphasizing on the fundamental shear strength and behavior of plate girders with trapezoidally corrugated webs (Fig. 4.1) used in bridge constructions (BGCWs) which are built with HSSs, with the main aim of combining the advantages of both the corrugated webs and the HSSs. This is based on the current authors' previous investigation [37], from which the description and validation of the current FE model could be revised.

4.5.1 Input data of the parametric studies

This subsection provides the details of the parametric studies generated to investigate the shear strength and behavior of BGCWs built with S460 HSSs [37]. The current parametric study has been conducted in BGCWs considering the original dimensions of bridge plate girders used in practical applications as presented in Table 3.1. However, to investigate the behavior of the current girders, variations in the web thickness (t_w) and height (h_w) have been undertaken. All girders were built from six corrugated waves and hence the length of the girders is $L = 6q$; where q is the length of one wave in the horizontal projection as represented in Fig. 4.1. The seven bridge corrugation dimensions, shown in Table 3.1, were considered in the developed FE models using the ABAQUS [20] computer program. This has resulted in 84 virtual bridge girders (i.e., 12 specimens for each bridge). In these girders, the width and thickness of the flanges, respectively, were fixed to 500 and 100 mm.

4.5.2 Shear deformations and stress distributions

The shear deformation of BGCWs built up with HSSs is discussed herein. Fig. 4.16 provides samples of the numerical outputs (captured in the elevation view) of the shear

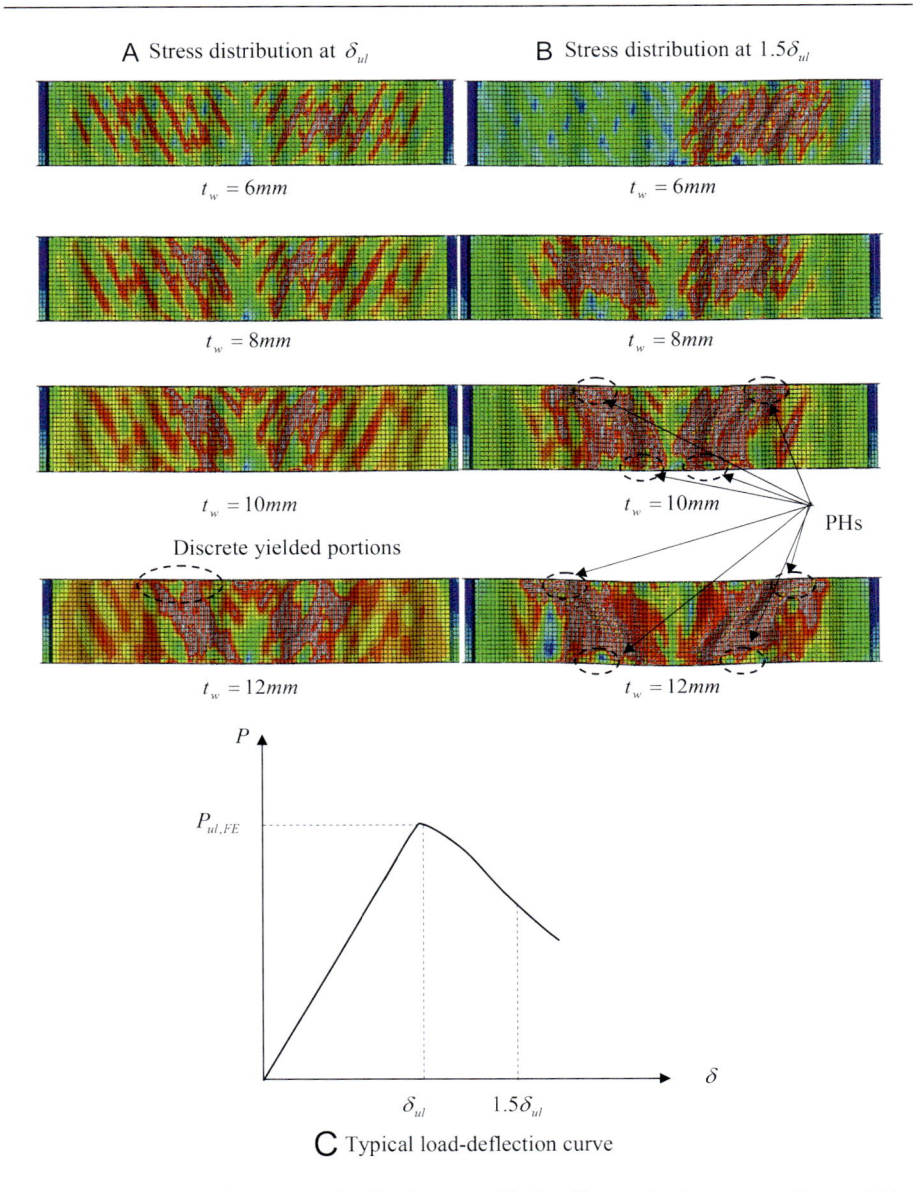

Figure 4.16 Typical shear stress distributions considering Cognac bridge corrugations at (A) δ_{ul} and (B) $1.5\delta_{ul}$ with definition of deflections shown in (C) for different web thicknesses (for interpretation of the references to color in this figure, the reader is referred to the web version of this chapter.).

deformation for cognac bridge, which has a web height (h_w) of 1500 mm, for the different considered web thicknesses (6, 8, 10, and 12 mm). To describe the shear deformations in different loading stages, this figure displays the Von Mises stress distribution at the deflection (δ_{ul}) corresponding to the ultimate load (in the left hand side of the figure) and at the descending branch of each girder at a deflection ($1.5\delta_{ul}$) (in the right hand side of the figure). In this figure, the red contour represents the regions which have just yielded (i.e., reached a stress of 460MPa), while the locations with grey contour are those regions exceeded the yield limit. Obviously, it can be noticed from Fig. 4.16 that increasing the web thickness leads to a significant increase in the yielded regions within the webs of the girders, which in turn raises the strengths of the girders.

On the other hand, Fig. 4.16A and B shows the shear failure mode of the cognac bridge with 6 mm and 8 mm web thicknesses, respectively. These girders do not show the formation of the shear plastic hinge (PH) because those girders failed in an interactive buckling mode; also shown in Table 4.12 as abbreviated by I. Accordingly, no differential deformation between the upper and lower flanges has been taken place to lead to the development of shear PHs [16,43]. On the contrary, in BGCWs with higher web thicknesses (10 mm and 12 mm) failing by global buckling (abbreviated by G in Table 4.11), the shear PHs appeared in the deformed shapes but at the descending parts of the load-deformation behavior. Fig. 4.17 is presented to show the vertical displacements of the upper and lower flanges at different cross-sections within the span of the girder having $t_w = 10$ mm, with the stress distribution shown in Fig. 4.16 as a sample. It is obvious that the displacements of both flanges are the same until the ultimate load is reached, for all sections considered, due to the significant out-of-plane stiffness of the corrugated webs. As the section becomes closer to the support, differences in the displacements between both flanges become obvious in the descending loading branch. In this loading stage, the web starts to undergo significant yielding. Hence, a reduction in the out-of-plane stiffness takes place so that the buckling resistance of the member becomes below its elastic value leading to the development of global buckling. After the development of the global buckling, the displacement of the upper flange becomes greater than that of the lower flange. Hence, the plastic hinge propagates. However, it is important to note that the girders at their ultimate loads (i.e. in the left side of Fig. 4.16) have not presented PHs once they had reached their capacities, which is in contrast to the behavior of IPGs as they fail at the onset of the formation of the PHs [43]. This may be attributed to uncompleted tension bands near the supporting flanges at the ultimate loads. Accordingly, the loss of the the out-of-plane stiffness still lower than the case of the continuous yielding (as in the descending parts of the load-deformation behavior—see the pictures to the right).

4.5.3 Ultimate shear strengths and comparisons with design models

4.5.3.1 Ultimate shear strengths

Herein, the ultimate shear strengths ($\tau_{ul,FE}$) and their comparisons with the available design models are presented, as can be seen in Table 4.12. It can easily be recognized

Table 4.12 Full details of analyzed models.

Bridge name	h_w [mm]	t_w [mm]	a [mm]	Failure mode	$\tau_{ul,FE}$ [MPa]	$\dfrac{\tau_{ul,FE}}{\tau_{n,Driver}}$	$\dfrac{\tau_{ul,FE}}{\tau_{n,Moon}}$	$\dfrac{\tau_{ul,FE}}{\tau_{n,Sause}}$	$\dfrac{\tau_{ul,FE}}{\tau_{n,Hassanien}}$	$\dfrac{\tau_{ul,FE}}{\tau_{n,Sause,m}}$	$\dfrac{\tau_{ul,FE}}{\tau_{n,Hassanien,m}}$
Shinkai	1183	6	2700	I	200	1.07	0.75	1.20	0.96	1.46	0.95
	1183	8		I	225	1.20	0.85	1.35	1.07	1.64	1.07
	1183	10		G	260	1.38	0.98	1.55	1.23	1.90	1.23
	1183	12		G	258	1.37	0.97	1.54	1.22	1.88	1.22
	1500	6		I	226	1.20	0.85	1.35	1.09	1.65	1.08
	1500	8		I	251	1.34	0.95	1.50	1.20	1.83	1.19
	1500	10		I	257	1.37	0.97	1.53	1.22	1.87	1.22
	1500	12		I	261	1.39	0.98	1.56	1.24	1.91	1.24
	2000	6		I	230	1.23	0.87	1.38	1.14	1.51	1.10
	2000	8		I	239	1.27	0.90	1.43	1.15	1.57	1.14
	2000	10		G	253	1.34	0.95	1.51	1.20	1.66	1.20
	2000	12		G	229	1.22	0.86	1.37	1.09	1.50	1.09
Matsnoki	2000	6	3360	I	160	0.85	0.60	0.96	0.82	1.17	0.77
	2000	8		I	178	0.95	0.67	1.06	0.87	1.30	0.85
	2000	10		G	201	1.07	0.75	1.20	0.96	1.46	0.95
	2000	12		G	220	1.17	0.83	1.31	1.05	1.61	1.04
	2210	6		G	210	1.12	0.79	1.25	1.09	1.37	1.01
	2210	8		G	191	1.02	0.72	1.14	0.93	1.25	0.91
	2210	10		G	229	1.22	0.86	1.37	1.10	1.50	1.09
	2210	12		G	202	1.07	0.76	1.21	0.96	1.32	0.96
	2500	6		I	175	0.93	0.66	1.05	0.93	1.02	0.84
	2500	8		G	213	1.13	0.80	1.27	1.05	1.24	1.01
	2500	10		G	204	1.09	0.77	1.22	0.98	1.19	0.97
	2500	12		G	180	0.96	0.68	1.08	0.86	1.05	0.85

(continued on next page)

Table 4.12 Full details of analyzed models—cont'd

Bridge name	h_w [mm]	t_w [mm]	a [mm]	Failure mode	$\tau_{ul,FE}$ [MPa]	$\dfrac{\tau_{ul,FE}}{\tau_{n,Driver}}$	$\dfrac{\tau_{ul,FE}}{\tau_{n,Moon}}$	$\dfrac{\tau_{ul,FE}}{\tau_{n,Sause}}$	$\dfrac{\tau_{ul,FE}}{\tau_{n,Hassanien}}$	$\dfrac{\tau_{ul,FE}}{\tau_{n,Sause,m}}$	$\dfrac{\tau_{ul,FE}}{\tau_{n,Hassanien,m}}$
Hondani	3000	6	3600	I	156	0.85	0.59	0.94	0.87	0.92	0.76
	3000	8		I	183	0.97	0.69	1.09	0.92	1.06	0.87
	3000	10		I	174	0.93	0.66	1.04	0.85	1.01	0.83
	3000	12		G	153	0.81	0.58	0.91	0.73	0.89	0.73
	3315	6		I	175	0.95	0.68	1.06	1.00	1.03	0.85
	3315	8		I	182	0.97	0.69	1.09	0.93	1.06	0.87
	3315	10		I	157	0.84	0.59	0.94	0.77	0.91	0.75
	3315	12		G	140	0.74	0.53	0.83	0.67	0.81	0.66
	3500	6		I	170	0.92	0.66	1.02	0.98	1.00	0.82
	3500	8		I	173	0.92	0.65	1.03	0.88	1.01	0.82
	3500	10		I	149	0.79	0.56	0.89	0.73	0.86	0.71
	3500	12		G	132	0.70	0.50	0.79	0.64	0.77	0.63
Cognac	1500	6	4032	I	159	0.89	0.60	0.97	0.83	1.20	0.78
	1500	8		I	182	0.97	0.69	1.09	0.89	1.33	0.87
	1500	10		G	201	1.07	0.76	1.20	0.96	1.47	0.95
	1500	12		G	215	1.15	0.81	1.29	1.03	1.57	1.02
	1771	6		I	150	0.83	0.56	0.92	0.80	1.13	0.73
	1771	8		I	179	0.95	0.67	1.07	0.88	1.31	0.86
	1771	10		G	202	1.07	0.76	1.21	0.97	1.47	0.96
	1771	12		G	205	1.09	0.77	1.23	0.98	1.50	0.98
	2000	6		I	142	0.79	0.53	0.87	0.77	1.07	0.70
	2000	8		I	167	0.89	0.63	1.00	0.83	1.22	0.80
	2000	10		G	193	1.03	0.72	1.15	0.93	1.40	0.92
	2000	12		G	202	1.07	0.76	1.21	0.96	1.47	0.96

(continued on next page)

Maupre	2400	6	3150	I	155	0.83	0.58	0.93	0.81	0.90	0.74
	2400	8		I	173	0.92	0.65	1.04	0.85	1.01	0.82
	2400	10		I	215	1.14	0.81	1.28	1.03	1.25	1.02
	2400	12		I	191	1.02	0.72	1.14	0.91	1.11	0.91
	2650	6		I	157	0.83	0.59	0.94	0.83	0.91	0.75
	2650	8		I	192	1.02	0.72	1.15	0.95	1.12	0.91
	2650	10		I	196	1.04	0.74	1.17	0.95	1.14	0.93
	2650	12		I	175	0.93	0.66	1.04	0.84	1.01	0.83
	3000	6		I	162	0.86	0.61	0.97	0.89	0.94	0.77
	3000	8		I	193	1.03	0.73	1.15	0.97	1.12	0.92
	3000	10		I	173	0.92	0.65	1.04	0.84	1.01	0.82
	3000	12		G	156	0.83	0.59	0.93	0.75	0.90	0.74
Dole	2300	6	4800	I	134	0.88	0.51	0.91	0.80	1.14	0.70
	2300	8		I	162	0.86	0.61	0.97	0.82	1.18	0.78
	2300	10		G	183	0.97	0.69	1.09	0.89	1.33	0.87
	2300	12		G	192	1.02	0.72	1.15	0.92	1.40	0.91
	2546	6		I	135	0.88	0.51	0.91	0.82	1.14	0.70
	2546	8		I	161	0.85	0.60	0.96	0.83	1.17	0.78
	2546	10		G	181	0.96	0.68	1.08	0.89	1.32	0.87
	2546	12		I	175	0.93	0.66	1.05	0.84	1.28	0.83
	2800	6		I	134	0.87	0.50	0.91	0.83	1.13	0.70
	2800	8		I	153	0.82	0.58	0.92	0.80	1.12	0.74
	2800	10		G	174	0.93	0.65	1.04	0.86	1.27	0.83
	2800	12		G	159	0.85	0.60	0.95	0.77	1.16	0.76

(*continued on next page*)

Table 4.12 Full details of analyzed models—cont'd

Bridge name	h_w [mm]	t_w [mm]	a [mm]	Failure mode	$\tau_{ul,FE}$ [MPa]	$\dfrac{\tau_{ul,FE}}{\tau_{n,Driver}}$	$\dfrac{\tau_{ul,FE}}{\tau_{n,Moon}}$	$\dfrac{\tau_{ul,FE}}{\tau_{n,Sause}}$	$\dfrac{\tau_{ul,FE}}{\tau_{n,Hassanien}}$	$\dfrac{\tau_{ul,FE}}{\tau_{n,Sause,m}}$	$\dfrac{\tau_{ul,FE}}{\tau_{n,Hassanien,m}}$
Ilsun	2000	6	3960	I	148	0.86	0.56	0.93	0.81	1.15	0.74
	2000	8		I	166	0.89	0.63	0.99	0.82	1.21	0.80
	2000	10		G	189	1.00	0.71	1.13	0.91	1.38	0.90
	2000	12		G	206	1.10	0.78	1.23	0.98	1.50	0.98
	2292	6		G	152	0.89	0.57	0.96	0.85	1.06	0.75
	2292	8		G	168	0.89	0.63	1.00	0.84	1.10	0.80
	2292	10		G	169	0.90	0.64	1.01	0.82	1.11	0.80
	2292	12		G	195	1.04	0.73	1.16	0.93	1.28	0.92
	2600	6		I	154	0.90	0.58	0.97	0.89	1.07	0.76
	2600	8		I	170	0.91	0.64	1.02	0.86	1.12	0.82
	2600	10		I	183	0.98	0.69	1.10	0.89	1.20	0.87
	2600	12		G	173	0.92	0.65	1.03	0.83	1.13	0.82
Ave						1.00	0.70	1.11	0.92	1.25	0.89
Standard deviation						0.157	0.119	0.181	0.130	0.262	0.144

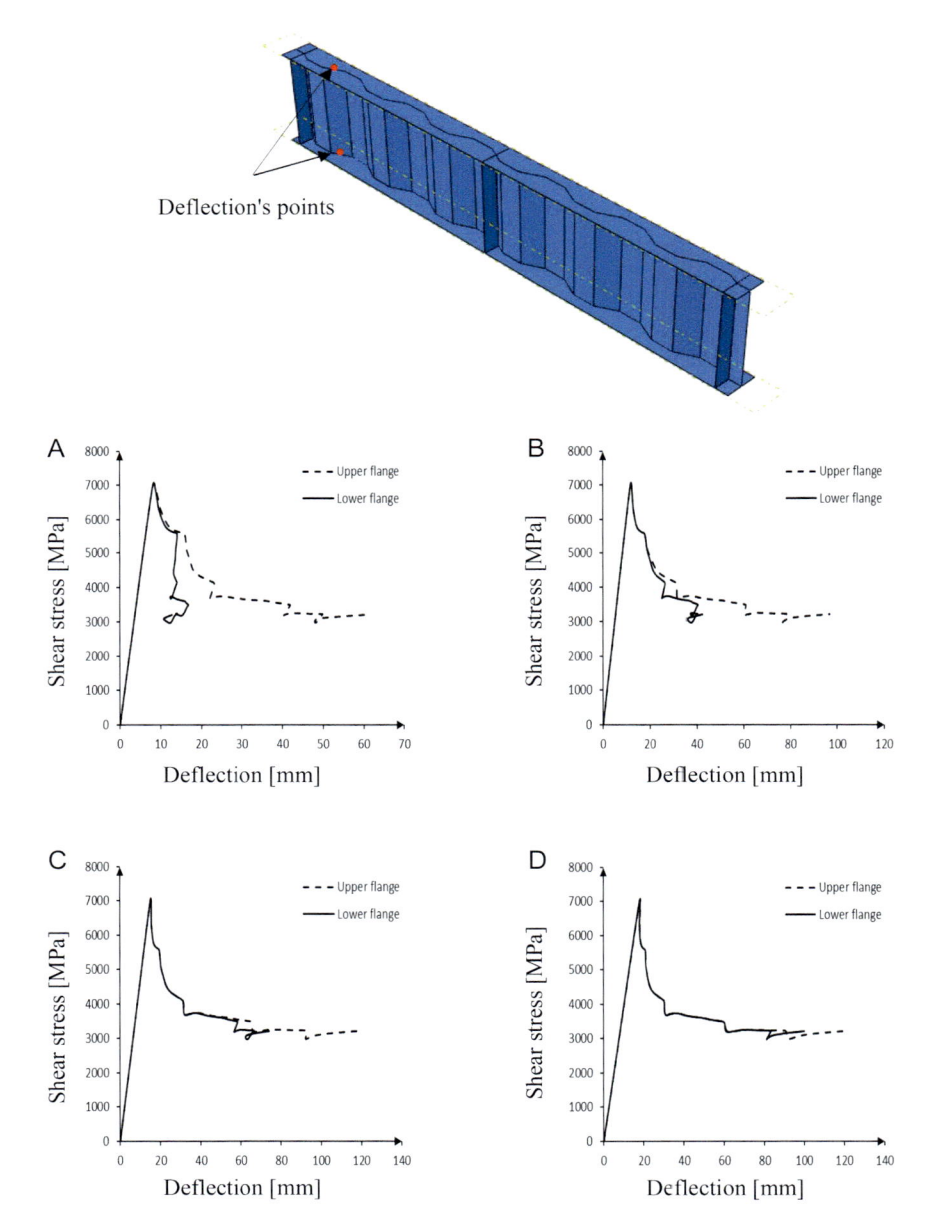

Figure 4.17 Load-deflection curves of upper and lower flanges at distances from the left support by: (A) one corrugation wave, (B) 1.5 corrugation waves, (C) 2 corrugation waves, and (D) 2.5 corrugation waves.

from the table that increasing the web thickness raises the shear strength of the girder irrespective of the corrugation dimensions. However, this increase in the $\tau_{ul,FE}$ becomes negligible and it even decreases if the girder started to fail globally when the thickness becomes relatively large (i.e., $t_w = 12$ mm). See for example the values of the $\tau_{ul,FE}$ for the girders formed from the Shinkai bridge corrugations with web thicknesses of 10 mm and 12 mm (in case of $h_w = 2000$ mm), from which it can be recognized that a decrease in the $\tau_{ul,FE}$ with about 10% (from 253 to 229 kN) resulted from an increase in the t_w of 20%. The reductions of the global ($\tau_{cr,G}$) shear buckling stress in such girders with large web thicknesses confirmed that $\tau_{cr,G} < \tau_{cr,L}$. Such deterioration in the girder's strength as a result of the global shear buckling has been discovered from early days. Hence, it has been recommended to design BGCWs by avoiding the occurrence of the global buckling which significantly reduces the strength of the girders [3]. Therefore, in practical designing of BGCWs, the values of $\tau_{cr,L}$ and $\tau_{cr,G}$ must first be examined for the condition $\tau_{cr,G} \geq \tau_{cr,L}$, which ensures that the failure mode is either a local or an interactive buckling mode before the final corrugation dimensions are confirmed.

4.5.3.2 Comparisons with design models

The comparison between the $\tau_{ul,FE}$ values and the available design models suggested by Driver et al. [3], Moon et al. [10], Sause and Braxtan [7], and Hassanein and Kharoob [17] is presented in this subsection. The details of these design models are provided earlier in the chapter. From the comparisons presented previously in Table 4.12, it can be noticed that the four design models provide large differences in their design shear strengths. As can be seen, the mean ratios of FE strengths-to-design strengths ($\tau_{ul,FE}/\tau_n$) are 1.00, 0.70, 1.11, and 0.92, respectively, using the design models of Driver et al. [3], Moon et al. [10], Sause and Braxtan [7], and Hassanein and Kharoob [17]. From Table 4.12, it can be noticed that the design model by Moon et al. [10] provides unsafe results for all different combinations of bridge dimensions, while others provide, in average, better strengths. However, despite that the average of the normalized values using the design model of Driver et al. [3] is equal to unity, this equation still unfavorable as much of the $\tau_{ul,FE}$ values were in the unsafe side. On the other hand, the design models by Sause and Braxtan [7] and Hassanein and Kharoob [17] predict the strengths of BGCWs much better. However, while the latter provides less scatter results, the former give the most suitable predictions. It is worth pointing out that the accuracy of design model by Hassanein and Kharoob [17] is less than that of Sause and Braxtan [14] because the former was initially proposed for girders with fixed junctures, as given previously in this chapter. Overall, the original design model suggested by Sause and Braxtan [7] is still found to be the best within the available models and may safely be used in designing BGCWs built with HSSs. This, however, has been recently confirmed by Leblouba et al. [38] for the girders formed from NSSs. This method is, however, restricted for specimens with initial imperfection values having the least of $h_w/200$ and t_w. In case the initial imperfection value was equal to the web thickness (t_w) as observed by Driver et al. [3] and was greater than $h_w/200$, the strength of the girder should be checked by FE analysis rather than by the suggested design model.

4.5.4 Effects of key parameters on the shear behavior of the bridge girders with corrugated webs

Based on the current parametric study, the effects of the considered parameters on the behavior of BGCWs are evaluated herein. Accordingly, this subsection presents and analyses the effects of the web thickness (t_w) and the web height (h_w). It is worth pointing out that the evaluation of each parameter is performed while the other parameters are kept constant.

4.5.4.1 Effects of web thickness

The effect of web thickness value is examined here using two bridge corrugation dimensions as samples of the qualitatively similar results. The relationships between the load and mid-span deflection are illustrated in Fig. 4.18. In Fig. 4.18A, the results of Cognac bridge with the web height of 1771 mm and shear span of 4032 mm (i.e., 3 corrugation waves) are presented, while Fig. 4.18B provides those of Ilsun bridge which has a web height and a shear span of 2000 mm and 3960 mm, respectively. From these results, it is obvious that increasing the web thickness increases subsequently the ultimate shear load. Additionally, it is noticed that increasing web thickness leads to a significant increase in the initial stiffness.

On the other hand, the efficiency of BGCWs is examined here by calculating the relative increase in the web cross-sectional area of each girder $(A_w/A_{w,6})$ as a result of increasing the web thickness with the increase in the ultimate shear strength; where $A_{w,6}$ is the web cross-sectional area of the girder with the least web thickness of 6 mm. As the areas of the upper flange, lower flange, and stiffeners were assumed constant for all girders, they were not included in the calculation of the cross-sectional areas. The efficiency is calculated herein for the girders with the four web thicknesses (i.e., 6, 8,10, and 12 mm) by comparing their ultimate strengths with the ultimate strength of the girder having the web of 6 mm thickness $(\tau_{ul,t_w}/\tau_{ul,6})$. Cognac bridge with web height equals 1771 mm is considered for this comparison because the results of other bridges are similar. From the comparison, it can be noticed that increasing the web thickness reduces the efficiency of the plate girder, as shown in Table 4.13. This table shows that increasing the web thickness leads to a significant increase in web's cross-sectional area against a relatively small increase in the ultimate strength. This observation means that using small web thicknesses is more economical than using large ones.

4.5.4.2 Effects of web height (h_w)

To investigate the effect of web height (h_w), Fig. 4.19 is provided to show the relationship between the load and the mid-span deflection for girders with three different h_w values. Those results are formed from the corrugations of Cognac bridge with web thickness (t_w) of 12 mm, as sample results, as the results of other corrugations are qualitatively similar. It can be noticed that increasing the web height (h_w) increases considerably the value of ultimate shear load as well as the initial stiffness. However, while the ultimate shear load increases with increasing the h_w value, the ultimate shear

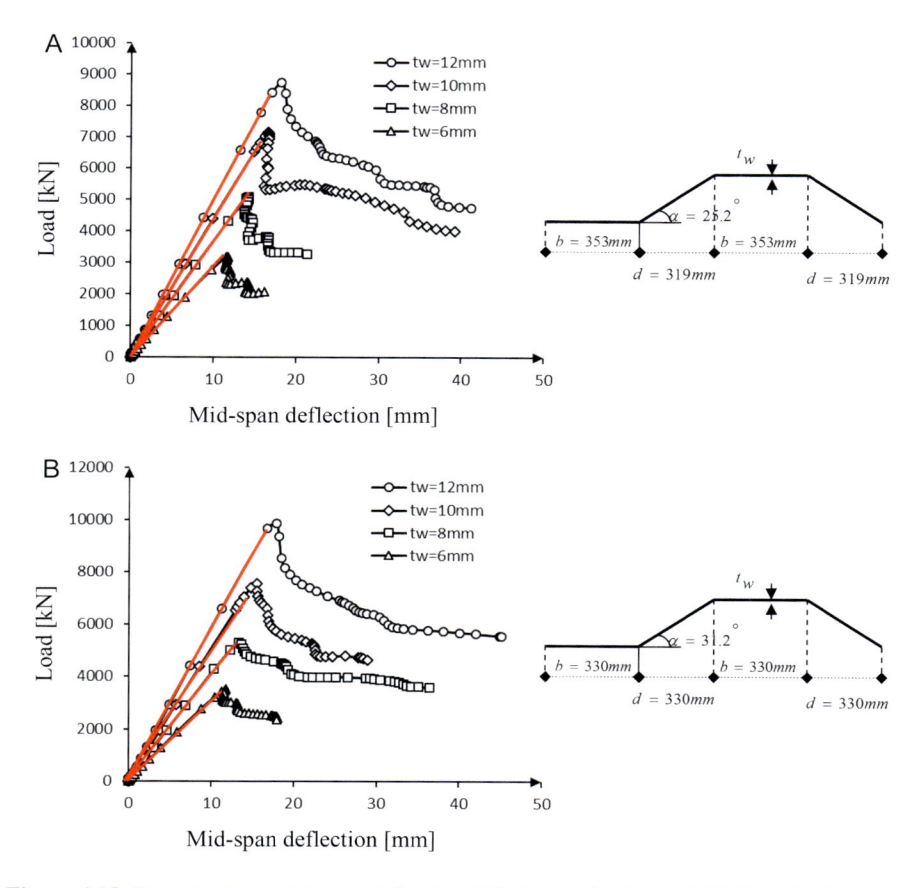

Figure 4.18 Shear load vs. mid-span deflection, (A) Cognac bridge and (B) Ilsun bridge (for interpretation of the references to color in this figure, the reader is referred to the web version of this chapter.).

Table 4.13 Efficiency of sample bridge girders with corrugated webs.

Thickness [mm]	A_w [mm^2]	τ_{ul} [MPa]	$\frac{A_w}{A_{w,6}}$	$\frac{\tau_{ul,tw}}{\tau_{ul,6}}$
6	10,626	299	1.00	1.00
8	14,168	358	1.33	1.19
10	17,710	404	1.67	1.35
12	21,252	411	2.00	1.38

stresses ($\tau_{ul,FE}$), shown in Table 4.12, decreases from 215 to 202 MPa. This means that the lower the h_w value, the more efficient the steel material is. On the other hand, it can be seen that smaller web heights tend to have a higher residual strength after the brittle failure of the girder; i.e. smaller web heights have a higher postbuckling strengths.

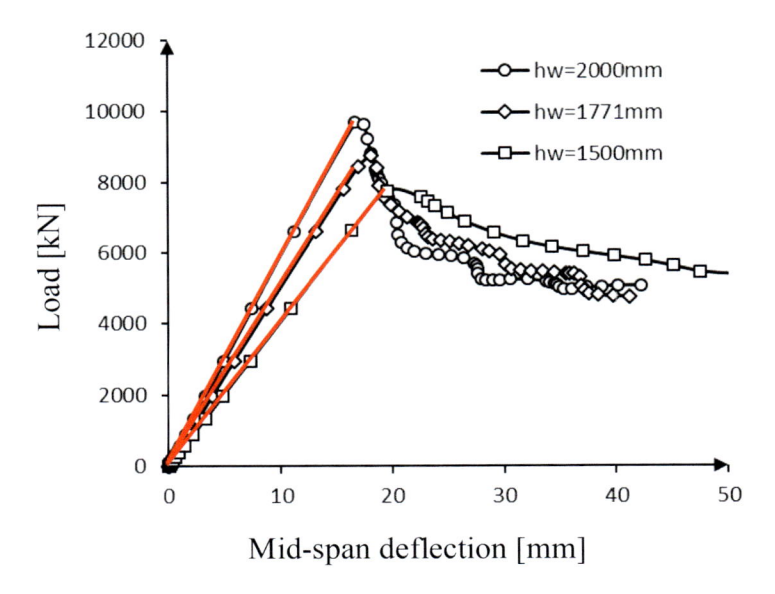

Mid-span deflection [mm]

Figure 4.19 Shear load vs. mid-span deflection for different Cognac BGCWs with $t_w = 12$ mm (for interpretation of the references to color in this figure, the reader is referred to the web version of this chapter.). *BGCWs*, bridge girders with corrugated webs.

4.5.4.3   Initial stiffness and deflection calculations

As can be seen in Figs 4.17 and 4.18, red straight lines were added to the elastic parts of the load-mid-span deflection relationships, from which it can be noticed that the linear elastic portions of the curves extend from the origin up to load levels very close to the ultimate loads. Accordingly, it might be possible to obtain the values of the maximum deflections of BGCWs with HSSs from the structural analyses based on the elastic deflection at the loaded point (Δ_e). This, accordingly, is checked in this subsection through the calculation of the initial stiffness (k_o), which could be quantified from the expression $P = k_o \Delta_e$ [39], where Eq.27.4 is the applied load. The initial stiffness is given by P:

$$k_o = \frac{k_b k_s}{k_b + k_s} \qquad (4.21)$$

where k_b is the bending stiffness given by Eq. (4.22). For the current FE models, the loads were applied at the mid-spans of the girders. Hence, the shear span is equal to half the overall span. Accordingly, k_b may instead be calculated from Eq. (4.23), where E is the elastic modulus, I_x is the moment of inertia of the section for overall girder bending (neglecting the web contribution due to the accordion effect [39] discussed earlier in

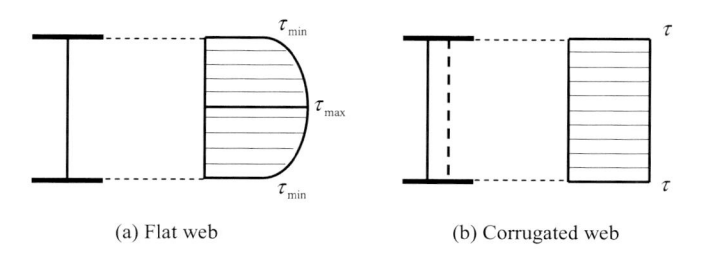

(a) Flat web   (b) Corrugated web

Figure 4.20 Distribution of the shear stress on the entire web.

this chapter), a and L are the shear span and the overall span, respectively.

$$k_b = \frac{3LEI_x}{a^2(L-a)^2} \tag{4.22}$$

$$k_b = \frac{48EI_x}{L^3} \tag{4.23}$$

On the other hand, the shear stiffness (k_s) of BGCWs can be calculated from the expression: $P = k_s\Delta_s$, where Δ_s is the shear deflection of the simply supported girder with concentrated load given by Eq. (4.24) [3].

$$\Delta_s = \frac{PL}{4AG}\alpha \tag{4.24}$$

where A is the area of the web ($h_w t_w$) which resists the shear force. The parameter (α) expresses the distribution of shear stress throughout the web of the section which is calculated as τ_{max}/τ_{ave}, where τ_{ave} is the average between τ_{max} and τ_{min}, as can be seen in Fig. 4.20A for the case of girders with flat webs. For the corrugated plates, the distribution of the shear force is constant on the entire web [3-4, 7]. So, α becomes equal to unity. So the equation calculating the shear stiffness for BGCWs can be deduced as given in Eqs. (4.25–4.28), where G_e represents the shear modulus [39] which considers the reduced shear stiffness owing to the existence of the web corrugations.

$$\Delta_s = \frac{PL}{4h_w t_w G_e} \tag{4.25}$$

$$k_s = \frac{P}{\Delta_s} \tag{4.26}$$

$$k_s = \frac{4h_w t_w}{L}G_e \tag{4.27}$$

$$G_e = \frac{E}{2(1+\upsilon)}\cdot\left(\frac{b+d}{b+c}\right) \tag{4.28}$$

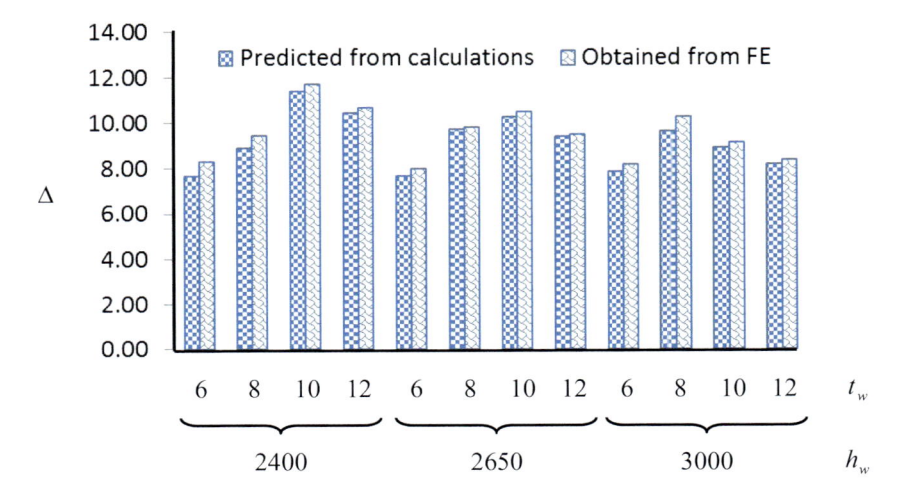

Figure 4.21 Comparison between predicted and FE deflections considering Maupré Bridge corrugations; all units in mm. *FE*, finite element.

As mentioned earlier, the deflection ($\Delta_{ul,Eq}$) at the ultimate loads is predicted considering that the elastic portion extends until the ultimate loads are reached and then compared with that obtained from FE analysis ($\Delta_{ul,FE}$). The comparison is shown here in Fig. 4.21 for sample results considering Maupré Bridge corrugations. From this figure, it can be seen that the above-mentioned suggestion may be valid because the results of both methods deviate slightly from each other, with an average ratio of about 0.97 (ranging between 0.93 and 0.99).

4.6 High-strength steel tapered girders

The behavior and strength of tapered corrugated web girders were investigated numerically by the current authors [40]. They used an elastic-perfect plastic stress-strain curve as it was found to simulate the real behavior of S460 HSSs. This accords with that recommended by Choi et al. [41], who found that the elastic-perfect plastic stress-strain curve of the HSSs provides safe results in design.

4.6.1 Parametric study

Since the data available in the literature is mostly based on tests of small-scale girders and the material thickness is generally much less than expected in practice, this part therefore uses the same dimensions of bridge plate girders used in practical application as given previously in Table 3.1. This parametric study [40] was generated by varying the flange inclination angle (γ^o) (i.e., 4, 8, 12, and 16°) and the thickness of the web

(t_w) (i.e., 6, 8, and 10 mm) in each typology. As a result, 333 FE models for TPGCWs were generated and analyzed (i.e., 51 specimens in each bridge except Shinkai and Cognac bridges which were used in 39 specimens). The width and thickness of the flanges, respectively, were 600 and 100 mm for all specimens. The Young's modulus (E) was taken as 200,000 MPa and Poisson's ratio (v) was used as 0.3. The steel used in these extended analyses was S460 which has yield and ultimate strengths of 460 MPa and 570 MPa, respectively, according to EN 1993-1-5 [19]. It should be noted that the current parametric studies are limited to girders that are failing by plastic and inelastic shear mechanisms. This is because the dimensions of the bridges, used in practice, do not fail elastically and, on the other hand, to make use of HSS material effectively [42]. This will appear at the end in the comparisons with design predictions.

4.6.2   Fundamental behavior

4.6.2.1   Ultimate shear strengths

Table 4.14 presents the ultimate strengths $(\tau_{ul,FE})$ of the nonlinear analysis. From that table, it can be observed that the ultimate shear strengths are affected by the girder typology. However, Cases I and II seem to have similar values, while the other cases are close to each other. To get deeper into the topic, comparisons of the ultimate shear strengths of Cases I and II on the one hand and Cases III and Cases IV on the other hand are provided in the same table. From this table, it can be seen that the average ratios of $\tau_{ul,FE,\ II}/\tau_{ul,FE,\ I}$ and $\tau_{ul,FE,\ III}/\tau_{ul,FE,\ IV}$ are 1.05 and 1.09, respectively, with standard deviations of 0.088 and 0.067. Hence, it can be concluded that the strengths of Cases I and II can be designed similarly to each other. In contrast, the design strengths of Cases III and Cases IV should be different.

4.6.2.2   Shear deformations and stress distributions

The shear stress carried by the flanges is first evaluated in this subsection, as shown in Fig. 4.22, which shows the distribution of the shear stress (S12) for a typical girder in MPa. As can be seen, the shear stress carried by the flanges is nearly zero. Hence, it might be neglected in design as nearly all of the shear force is carried by the web. This is similar to previous available findings on the accordion effect of corrugated web girders [1–3].

On the other hand, typical shear stress distributions for sample results of the four typologies, close to the ultimate load of each girder, are provided in Fig. 4.23. The Cognac Bridge is considered for these results because the results of other bridges are qualitatively similar. Specifically, the figure shows those girders with an inclination angle of 8°. It can be seen that the shear failure in all cases is concentrated near the shorter height of the tapered web where the shear stress is maximum.

Also, it can be noticed (from Table 4.14) that Cases II and III have higher ultimate shear strengths as compared to Cases I and IV, respectively. This may be attributed to relationship between the direction of the propagated tension field (presented by dotted

Table 4.14 Details and stresses of tapered S460 HSS models.

Bridge name	h_{w1} [mm]	$\gamma°$ [°]	h_{wo} [mm]	$t_w{*}$ [mm]	a [mm]	Prismatic	$\tau_{ul,FE}$ [MPa] Case I	Case II	Case III	Case IV	$\frac{\tau_{ul,FE,II}}{\tau_{ul,FE,I}}$	$\frac{\tau_{ul,FE,III}}{\tau_{ul,FE,IV}}$
Shinkai	1183	4	931	6	3600	201	225	255	193	181	1.13	1.07
		8	677				327	313	189	175	0.96	1.15
		12	418				472	462	183	174	0.98	1.05
		4	931	8		225	244	258	197	196	1.06	1.01
		8	677				348	330	191	178	0.95	1.07
		12	418				501	490	185	174	0.98	1.06
		4	931	10		237	264	264	216	211	1.00	1.02
		8	677				355	351	188	187	0.99	1.01
		12	418				507	503	188	174	0.99	1.08
Matsnoki	2210	4	1897	6	4480	157	166	181	155	157	1.09	0.99
		8	1581				203	188	144	136	0.93	1.06
		12	1258				303	232	140	136	0.77	1.03
		16	926				305	318	140	135	1.04	1.04
		4	1897	8		194	197	192	190	164	0.97	1.16
		8	1581				215	213	158	154	0.99	1.03
		12	1258				308	251	156	152	0.81	1.03
		16	926				321	341	156	144	1.06	1.08
		4	1897	10		211	226	224	198	191	0.99	1.04
		8	1581				233	248	179	174	1.06	1.03
		12	1258				318	264	167	158	0.83	1.06
		16	926				335	358	165	152	1.07	1.09

(continued on next page)

Hondani	3315	4	2980	6	4800	161	163	164	171	143	1.01	1.20
		8	2641				166	181	142	131	1.09	1.08
		12	2295				178	197	136	119	1.11	1.14
		16	1939				191	219	132	116	1.15	1.14
		4	2980	8		191	192	196	201	170	1.02	1.18
		8	2641				195	205	170	148	1.05	1.15
		12	2295				198	219	146	129	1.11	1.13
		16	1939				210	236	139	129	1.12	1.08
		4	2980	10		178	195	198	179	180	1.02	0.99
		8	2641				197	221	157	160	1.12	0.98
		12	2295				213	238	157	150	1.12	1.05
		16	1939				219	255	141	130	1.16	1.08
Cognac	1771	4	1395	6	5375	149	173	171	144	139	0.99	1.04
		8	1016				209	235	136	130	1.12	1.05
		12	629				329	368	138	128	1.12	1.08
		4	1395	8		181	195	196	169	155	1.01	1.09
		8	1016				240	262	164	145	1.09	1.13
		12	629				358	417	155	143	1.16	1.08
		4	1395	10		197	209	220	194	172	1.05	1.13
		8	1016				252	277	175	155	1.10	1.13
		12	629				376	423	165	150	1.13	1.10

(continued on next page)

Table 4.14 Details and stresses of tapered S460 HSS models—cont'd

| Bridge name | h_{w1} [mm] | $\gamma^o[°]$ | h_{wo} [mm] | $t_{w\cdot}$ [mm] | a [mm] | $\tau_{ul,FE}$ [MPa] | | | | | $\frac{\tau_{ul,FE,II}}{\tau_{ul,FE,I}}$ | $\frac{\tau_{ul,FE,III}}{\tau_{ul,FE,IV}}$ |
						Prismatic	Case I	Case II	Case III	Case IV		
Maupre	2650	4	2356	6	4200	159	160	176	168	139	1.10	1.21
		8	2060				169	192	163	134	1.14	1.22
		12	1758				196	198	150	134	1.01	1.12
		16	1446				218	241	146	131	1.11	1.11
		4	2356	8		204	182	218	184	175	1.20	1.05
		8	2060				191	221	173	163	1.16	1.06
		12	1758				202	259	159	149	1.28	1.07
		16	1446				237	264	151	139	1.11	1.09
		4	2356	10		212	192	231	185	193	1.20	0.96
		8	2060				204	229	183	184	1.12	0.99
		12	1758				218	263	173	171	1.21	1.01
		16	1446				260	278	163	158	1.07	1.03
Dole	2546	4	2099	6	6400	133	159	149	137	131	0.94	1.05
		8	1647				182	179	130	111	0.98	1.17
		12	1186				208	243	126	103	1.17	1.22
		16	712				360	400	117	101	1.11	1.16
		4	2099	8		159	170	175	153	159	1.03	0.96
		8	1647				198	205	147	138	1.04	1.07
		12	1186				247	271	138	116	1.10	1.19
		16	712				383	427	127	109	1.11	1.17
		4	2099	10		184	190	194	184	176	1.02	1.05
		8	1647				214	221	156	138	1.03	1.13
		12	1186				271	275	147	123	1.01	1.20
		16	712				402	464	136	113	1.15	1.20

(continued on next page)

Ilsun	2292	4	1923	6	5280	145	177	161	135	144	0.91	0.94
		8	1550				193	198	130	121	1.03	1.07
		12	1170				240	241	126	118	1.00	1.07
		16	779				332	356	120	109	1.07	1.10
		4	1923	8		163	186	186	173	165	1.00	1.05
		8	1550				220	214	161	138	0.97	1.17
		12	1170				256	260	147	130	1.02	1.13
		16	779				352	380	139	123	1.08	1.13
		4	1923	10		179	189	204	184	188	1.08	0.98
		8	1550				225	231	176	149	1.03	1.18
		12	1170				267	273	159	140	1.02	1.14
		16	779				368	392	147	138	1.07	1.07
Ave											1.05	1.09
SD											0.088	0.067

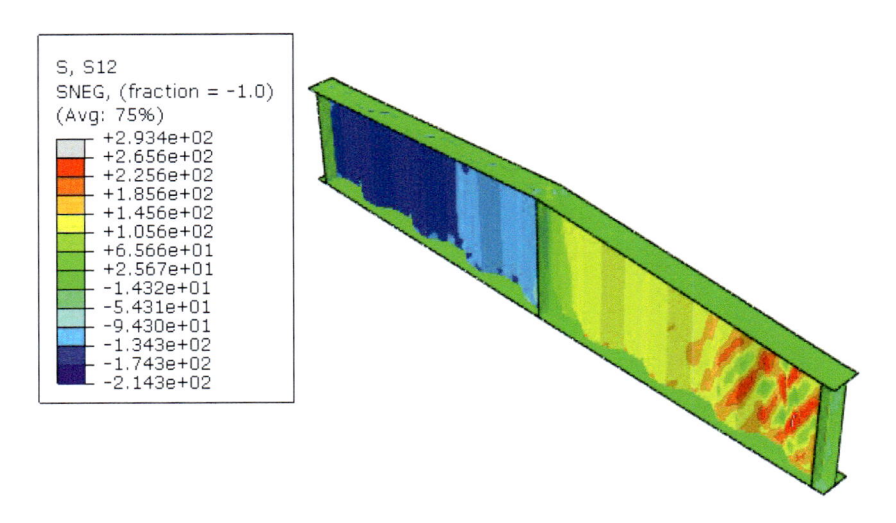

Figure 4.22 Typical distribution of stress S12 in TPGCWs in Case I [in MPa].

arrows in Fig. 4.23) and the type of force in the inclined flange (shown by solid arrows in the same figure). In Cases II and III, the direction of tension field appears in the same direction of the force in the inclined flange; tension in the lower inclined flange in Case II, and compression in the upper inclined flange in Case III. In contrast, the propagated tension fields in cases I and IV oppositely counteract the forces available in the inclined flanges; see Fig. 4.23. Subsequently, it can be concluded that when the direction of the force in the inclined flange has the same direction of the tension field, the ultimate shear strength increases.

On the other hand, the failure mode of TPGCWs was found to be different from that of TPGFWs. The latter, as shown in the research of Bedynek et al. [18], is characterized by the development of shear plastic hinges (SPHs) at their top and bottom flanges. Conversely, such plastic hinges have not developed in the current TPGCWs with bridge dimensions. The development of SPHs in plate girders with flat flanges is attributed [43] to the differential shear deformation between the top and bottom flanges, as shown in Fig. 4.24 which represents the deformed shape of one of the validated girders tested by Bedynek et al. [18]. When the flat web plate buckles, the top flange displaces in the vertical direction greater than that of the bottom. Hence, it was important to examine the differential shear deformation between the top and bottom flanges in current TPGCWs, as can be seen in Fig. 4.25, which provides the vertical deflection of several points on both flanges of the girder presented in Fig. 4.23A, as an sample. From this figure, it can be seen that the deflection of the upper and lower flanges is almost typical in every vertical section near the support, due to the significant out-of-plane stiffness of the corrugated webs. This is because the out-of-plane deformation of the corrugated webs is relatively very small or does not occur. So, no relative shear deformation occurs between the flanges to form SPHs. However, this conclusion is valid just for the current TPGCWs with bridge dimensions, where

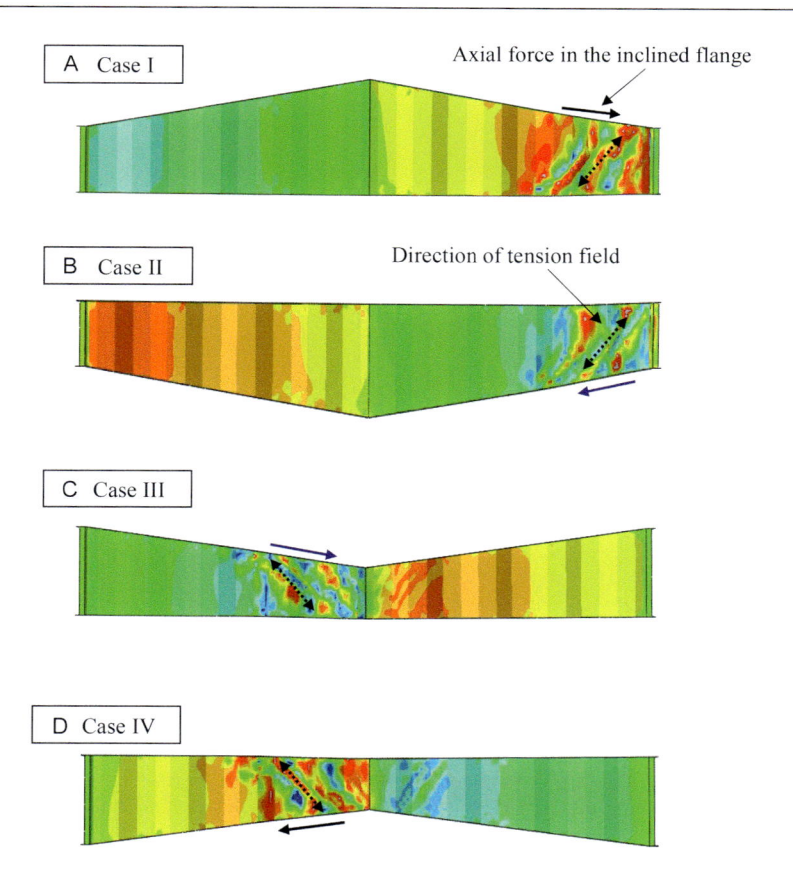

Figure 4.23 Typical shear stress distributions considering Cognac bridge.

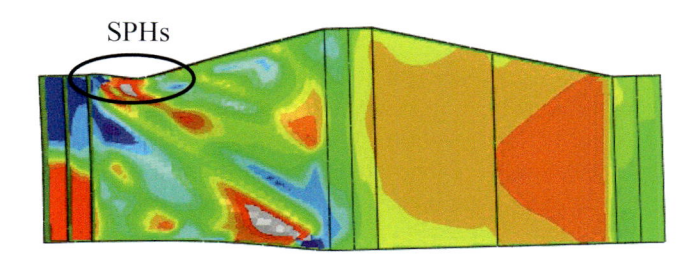

Figure 4.24 Shear plastic hinge in tapered plate girder with flat web.

the failure takes place by the interactive buckling. Hence, it is expected that SPHs may occur in those girders failing by global buckling. This, however, should be checked for girders with smaller corrugation dimensions compared to those used for bridges in further publications.

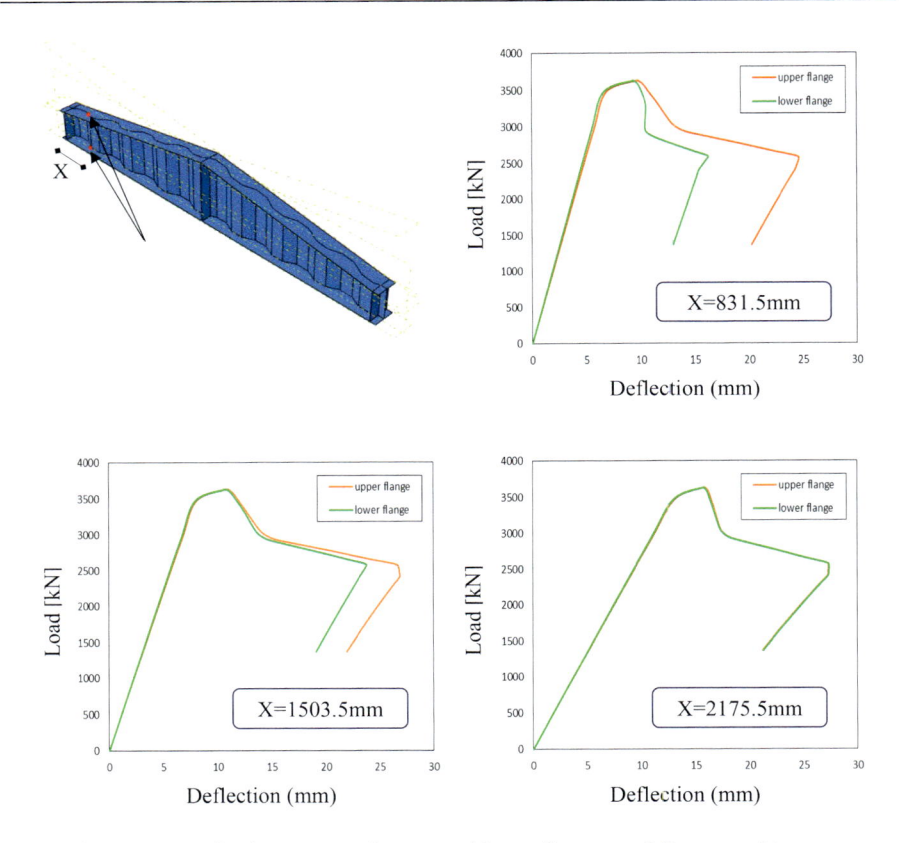

Figure 4.25 Load-deflection curves of upper and lower flanges at different positions.

4.6.3 Design strength

4.6.3.1 Comparisons with available design strength model

Hassanein and Kharoob [44] suggested an equation to predict the ultimate shear strength ($\tau_{ul,D}$) for different typologies which is presented here by Eq. (4.29), based on the design manual for PC bridges with corrugated steel webs [13] with modification by applying the coefficient c_T. This coefficient, called as the tapering coefficient, is assumed unity in typologies I and II and is taken as the ratio between the short-to-long depths of the tapered webs (h_{wo}/h_{w1}) in typologies III and IV. The current FE strengths of different bridge girder models are compared with the design model proposed by Hassanein and Kharoob [44] in Fig. 4.26, which presents the relationships between normalized strengths ($\tau_{ul,FE}/\tau_y$) versus the slenderness ratios (λ_s). The slenderness ratios (λ_s) used in the current calculations of the data points shown in these figures were determined based on the critical shear stresses proposed previously by Hassanein et al. [37]. From this figure, it can be observed that the design model provides substantial unsafe results. This may be attributed to the fact that the design model by Hassanein and Kharoob [44] was suggested for girders formed from normal mild steels only with

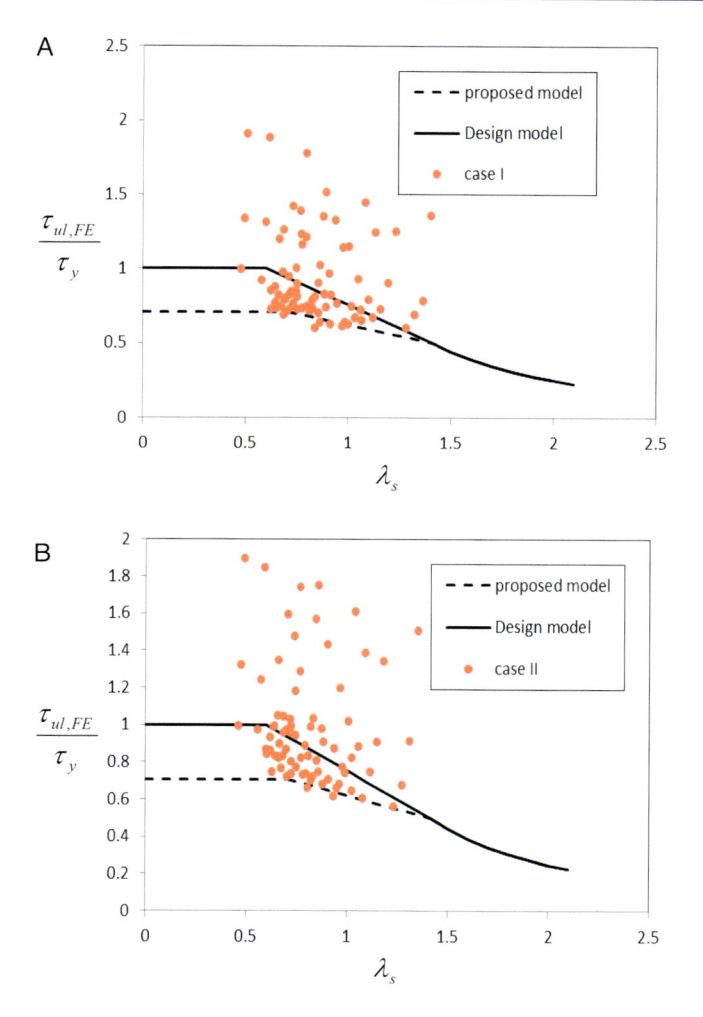

Figure 4.26 Proposed model versus slenderness parameter; (A) Case I, (B) Case II, (C) Case III, and (D) Case IV.

a length of two corrugation dimensions. Hence, an enhanced ultimate shear buckling strength for each typology is suggested in the following subsection.

$$\frac{\tau_{ul,D}}{\tau_y} = c_T \begin{cases} 1.0 & : \lambda_s \leq 0.6 \\ 1 - 0.614(\lambda_s - 0.6) & : 0.6 < \lambda_s \leq \sqrt{2} \\ \frac{1}{\lambda_s^2} & : \sqrt{2} < \lambda_s \end{cases} \tag{4.29}$$

$$\lambda_s = \sqrt{\frac{\tau_y}{\tau_{cr,T}}} \tag{4.30}$$

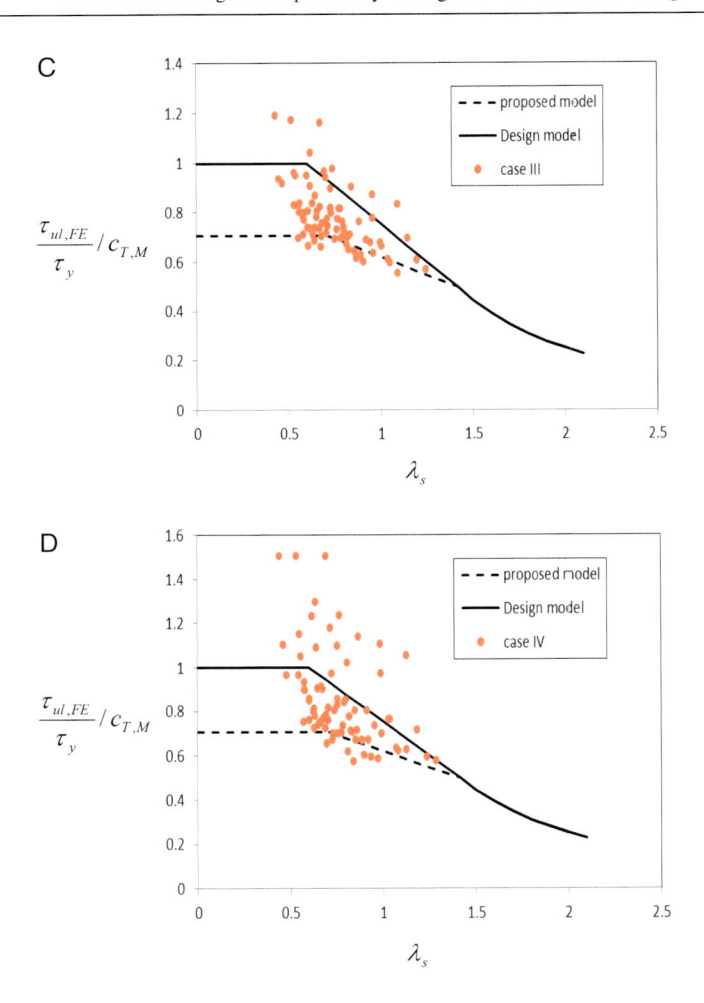

Figure 4.26, cont'd.

4.6.3.2   Suggested design equation of TPGCWs

In this subsection, an equation is suggested to calculate the ultimate shear strength of different typologies of TPGCWs formed from HSSs, with the main objective of providing a lower bound for FE strengths only in the plastic and inelastic stages. This is because the current investigation did not account for girders that failed elastically to take advantage of the high strengths of the HSSs. Eq. (4.37) presents this proposed equation. The boundary between the first two stages of λ_s was additionally modified. As can be noticed, the normalized strength of the first stage is deceased to 0.707 with λ_s range extended to 0.707 instead of 0.6. Fig. 4.26 shows that all FE strengths become conservatively predicted by the current proposals for all typologies. The point to note is that the tapering coefficient ($c_{T,M}$) has now two different values for Cases III and

IV based on the strength results shown in Subsection 4.6.2.1, from which it is shown that the design strengths of Cases III and Cases IV should be different. It should be mentioned that the very high conservative strength results have been occurred in girders with large values of the flange inclination angles which can rarely be used in practice.

$$\frac{\tau_{ul,prop}}{\tau_y} = c_{T,M}\begin{cases} 0.707 & : \lambda_s \leq 0.707 \\ 0.707 - 0.293\,(\lambda_s - 0.707): 0.707 < \lambda_s \leq \sqrt{2} \\ \frac{1}{\lambda_s^2} & :\sqrt{2} < \lambda_s \end{cases} \tag{4.31}$$

$$c_{T,M} = \begin{cases} 1.00 & caseI,\ caseII \\ (h_{wo}/h_{w1})^{0.4} & caseIII \\ (h_{wo}/h_{w1})^{0.8} & caseIV \end{cases} \tag{4.32}$$

References

[1]   R.W. Hamilton, "Behavior of welded girder with corrugated webs", Ph.D. thesis, University of Maine, Orono, Maine, 1993.

[2]   M. Elgaaly, R.W. Hamilton, A. Seshadri, Shear strength of beams with corrugated webs, J. Struct. Eng. 122 (4) (1996) 390–398.

[3]   R.G. Driver, H.H. Abbas, R. Sause, Shear behavior of corrugated web bridge girders, J. Struct. Eng., ASCE 132 (2) (2006) 195–203.

[4]   J. Yi, H. Gil, K. Youm, H. Lee, Interactive shear buckling behavior of trapezoidally corrugated steel webs, Eng. Struct. 30 (2008) 1659–1666.

[5]   A.S. El-Metwally "Prestressed composite girders with corrugated steel webs", M.S. thesis. Calgary, AB, Department of Civil Engineering, University of Calgary; 1998.

[6]   E.Y. Sayed-Ahmed, Behavior of Steel and (or) composite girders with corrugated steel webs, Can. J. Civ. Eng. 28 (4) (2001) 656–672.

[7]   R. Sause, T.N. Braxtan, Shear strength of trapezoidal corrugated steel webs, J. Constr. Steel Res. 67 (2011) 223–236.

[8]   J. Moon, J.-W. Yi, B.H. Choi, H-E. Lee, Lateral-torsional buckling of I-Girder with corrugated webs under uniform bending, Thin-Walled Struct. 47 (2009) 21–30.

[9]   M.F. Hassanein and O.F. Kharoob, "Behavior of bridge girders with corrugated webs: (I) real boundary conditions at the juncture of the web and flanges", Eng. Struct., DOI: 10.1016/j.engstruct.2013.03.004, 2013.

[10] J. Moon, J. Yi, B.H. Choi, H. Lee, Shear strength and design of trapezoidally corrugated steel webs, J. Constr. Steel Res. 65 (2009) 1198–1205.

[11] H. Gil, S. Lee, J. Lee, H. Lee, Shear buckling strength of trapezoidally corrugated steel webs for bridges, Transport. Res. Record: J. Transport. Res. Board (2005) 473–480 CD 11-S.

[12] M.F. Hassanein, O.F. Kharoob, Shear strength and behavior of transversely stiffened tubular flange plate girders, Eng. Struct. 32 (2010) 2617–2630.

[13] JSCE (Japan Society of Civil Engineers). Design Manual for PC Bridges with Corrugated Steel Webs. Research Committee for Hybrid Structures with Corrugated Steel Webs, Japan Society of Civil Engineers, Tokyo, Japan, 1998.

[14]   H.H. Abbas, "Analysis and design of corrugated web I-girders for bridges using high performance steel", Ph.D. dissertation, Bethlehem, PA, Department of Civil and Environmental Engineering, Lehigh University, 2003.

[15]   J.T. Easley, Buckling formulas for corrugated metal shear diaphragms, J. Struct. Div., SECF ST7 (1975) 1403–1417.

[16]   J. Lindner, B. Huang, Beulwerte für trapezförmig profilierte bleche unter schubbeanspruchung, Stahlbau 64 (2) (1995) 370–374.

[17]   M.F. Hassanein, O.F. Kharoob, Behavior of bridge girders with corrugated webs: (II) shear strength and design, Eng. Struct. 57 (2013) 544–553.

[18]   A. Bedynek, E. Real, E. Mirambell, Tapered plate girders under shear: tests and numerical research, Eng. Struct. 46 (2013) 350–358.

[19]   EN 1993-1-5, Eurocode 3: design of steel structures - Part 1-5: Plated structural elements, CEN, Brussels, Belgium, 2007.

[20]   ABAQUS ABAQUS Standard User's Manual The Abaqus Software is a product of Dassault Systèmes Simulia Corp., Providence, RI, USA Dassault Systèmes, Version 6.8, 2008.

[21]   J.-G. Nie, L. Zhu, M.-X. Tao, L. Tang, Shear strength of trapezoidal corrugated steel webs, J. Constr. Steel Res. 85 (2013) 105–115.

[22]   J. Yi, H. Gil, K. Youm, H. Lee, Interactive shear buckling behavior of trapezoidally corrugated steel webs, Eng. Struct. 30 (2008) 1659–1666.

[23]   H.H. Abbas, R. Sause, R.G. Driver, Shear strength and stability of high performance steel corrugated web girders, in: SSRC Conference, 2002, p. 361. -187.

[24]   EN 1993-1-1, Eurocode 3: design of steel structurespart 1-1: general rules and rules for buildings, CEN, Brussels, Belgium, 2004.

[25]   B. Lai, J.Y.R. Liew, A.L. Hoang, Behavior of high strength concrete encased steel composite stub columns with C130 concrete and S690 Steel, Eng. Struct. 200 (2019) 109743.

[26]   Y. Sun, Y. Liang, O. Zhao, Testing, numerical modelling and design of S690 high strength steel welded I-section stub columns, J. Constr. Steel Res. 159 (2019) 521–533.

[27]   M. Gkantou, M. Theofanous, C. Baniotopoulos, Plastic design of hot-finished high strength steel continuous beams, Thin-Walled Struct. 133 (2018) 85–95.

[28]   L. Zhang, F. Wang, Y. Liang, O. Zhao, Experimental and numerical studies of press-braked S690 high strength steel channel section beams, Thin-Walled Struct. 148 (2020) 106499.

[29]   H. Fang, T.-M. Chan, Buckling resistance of welded high-strength-steel box-section members under combined compression and bending, J. Constr. Steel Res. 162 (2019) 105711.

[30]   E. Gogou, "Use of high strength steel grades for economical bridge design", Master Thesis Study, Delft University of Technology Iv-Infra, Amsterdam, 2012.

[31]   D.M. West, C. Lansang, "Global manufacturing scorecard: How the US Compares to 18 Other Nations", Research report by The Brookings Institution, 2018 (https://www.brookings.edu/research/global-manufacturing-scorecard-how-the-us-compares-to-18-other-nations/).

[32]   G.-Q. Li, H. Lyu, C. Zhang, Post-fire mechanical properties of high strength Q690 structural steel, J. Constr. Steel Res. 132 (2017) 108–116.

[33]   P.V. Nidheesh, M.S. Kumar, An overview of environmental sustainability in cement and steel production, J. Cleaner Prod. 231 (2019) 856–871.

[34]   WSA: World Steel AssociationThe Three Rs of Sustainable Steel, World Steel Association, Brussels, Belgium, 2010.

[35] WSA: World Steel AssociationClimate change and the production of iron and steel, World Steel Association, Brussels, Belgium, 2021.

[36] The Paris Agreement, United Nations Framework Convention on Climate Change, Process and meetings, https://unfccc.int/process-and-meetings/the-paris-agreement/the-paris-agreement.

[37] M.F. Hassanein, A.A. Elkawas, A.M. El Hadidy, M. Elchalakani, Shear analysis and design of high-strength steel corrugated web girders for bridge design, Eng. Struct. 146 (2017) 18–33.

[38] M. Leblouba, M.T. Junaid, S. Barakat, S. Altoubat, M. Maalej, Shear buckling and stress distribution in trapezoidal web corrugated steel beams, Thin-Walled Struct. 113 (2017) 13–26.

[39] R.P. Johnson, J. Cafolla, Corrugated webs in plate girders for bridges, Proc. Inst. Civil Eng.—Struct. Build. 123 (1997) 157–164.

[40] A.A. Elkawas, M.F. Hassanein, M.H. El-Boghdadi, Numerical investigation on the non-linear shear behaviour of high-strength steel tapered corrugated web bridge girders, Eng. Struct. 134 (2017) 358–375.

[41] Y.S. Choi, D. Kim, S.C. Lee, Ultimate shear behaviour of web panels of HSB800 plate girders, Construct. Build. Mater. 101 (2015) 828–837.

[42] C.-S. Wu, M.F. Hassanein, H. Deng, Y.-M. Zhang, Y.-B. Shao, Shear buckling response of S690 steel plate girders with corrugated webs, Thin-Walled Struct. 157 (2020) 107015.

[43] M.M. Alinia, M. Shakiba, H.R. Habashi, Shear failure characteristics of steel plate girders, Thin-Walled Struct. 47 (2009) 1498–1506.

[44] M.F. Hassanein, O.F. Kharoob. Shear buckling Behavior of Tapered Bridge Girders with Steel Corrugated Webs, Eng. Struct., 74 (2014) pp. 157–169.

Flexural buckling behavior

5.1 General

I-section plate girders with flat webs (IPGs) are currently the main structural elements in many constructional areas [1]. They are mainly designed for bending moments (BMs). So, they are generally formed from cross-sections of large depths and slender web plates. These slender web plates reduce the flexural strengths of the girders because of their weak out-of-plane bending stiffness. To avoid such weak out-of-plane bending stiffness of the flat web plates, transverse stiffeners are often used in IPGs. This, however, causes a significant increase in the girder's weight and increases the risk of fatigue failure due to the welds used to connect the stiffeners to the flanges and the web. Recently, the use of corrugated webs has been found to raise the bending stiffness in the lateral direction of the girder without the use of stiffeners, resulting in much lower weight and higher fatigue resistance [2]. Accordingly, corrugated web girders (CWGs) have been used in bridges worldwide [3]. However, these girders were designed based on half-scale experiments [2]. To take advantage of these innovative girders, design models for different straining actions must be provided. Despite that, the literature survey [4] shows that most studies on the behavior of CWGs have been conducted to study the shear, while the lateral-torsional buckling (LTB) behavior has attracted fewer investigations [5–11].

In the absence of sufficient lateral bracing of the compression flange of the girder, the plastic section moment strength of the girder cannot be developed. Hence, LTB becomes the main failure mode of practical steel I-section girders under the action of in-plane bending. This is because it considerably reduces the strength of such girders relative to their cross-sectional resistances. This LTB is a superposition of the flexural and torsional buckling modes [12]. The flexural buckling mode of doubly symmetric cross-sections (Fig. 5.1A) mainly occurs under axial compression, and it is characterized by the in-plane or out-of-plane displacement (u) of the cross-section of the girder. As can be noticed from the figure, the cross-sectional displacement in flexural buckling is not accompanied by twisting. As a common fact, the current design of members undertaking flexural buckling is based on the critical buckling load derived by Euler [13] in 1759. Conversely, the torsional buckling leads to a cross-sectional twist (ϕ) that takes place along the length of the member without any lateral displacement. Basically, this buckling type was explored by Saint-Venant [14], in 1855, who found that the original and buckled configurations of the member coincide, as can be observed from Fig. 5.1B. This buckling type occurs in columns composed of crusade sections under axial compression loading and in I-section beams under torsion. However, in LTB, the cross-section of the member undergoes both lateral displacements (u) and twists (ϕ), as shown in Fig. 5.1C. In 1899, Michell [15] and Prandtl [16] independently explored thin rectangular beams under transverse loading. Based on their results, they derived a quadratic differential equation in both u and ϕ [17]. About

Behavior and Design of Trapezoidally Corrugated Web Girders for Bridge Construction: Recent Advances.
DOI: https://doi.org/10.1016/B978-0-323-88437-2.00002-2

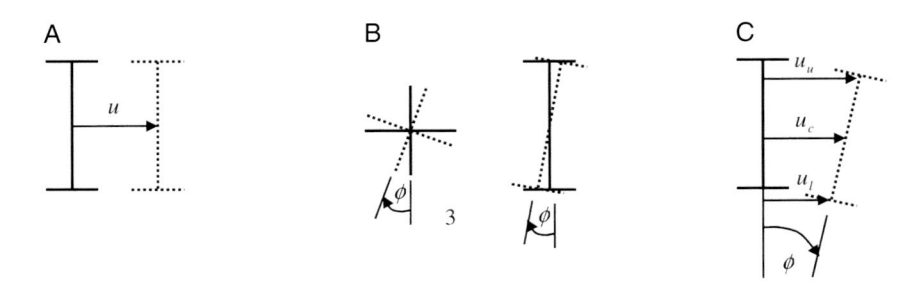

Figure 5.1 Global buckling types of structural steel members—solid cross-sections represent the original shapes, while dotted ones are the buckled shapes.

40 years later, Vlasov [18] was the first to introduce the elastic LTB moments of simply-supported I-section girders under the worst case of loading (i.e., the constant BM along the girder's span). Then, Timoshenko [19–20] revealed that the flexural behavior and strength of thin-walled open I-section members is controlled by *warping torsion* [21]. Consequently, a formula (Eq. 5.1) for the elastic critical LTB moment ($M_{cr,LTB}$) was derived [12] for simply supported I-section girder under constant BM.

$$M_{cr,LTB} = \frac{\pi}{L}\sqrt{EI_y GJ\left(1 + \frac{\pi^2}{L^2}\frac{EC_w}{GJ}\right)} \tag{5.1}$$

where

L is the unbraced length, EI_y is the minor-axis flexural rigidity, GJ is the torsional rigidity, and EC_w is the warping rigidity given by:

$$GJ = \frac{E}{2(1+\upsilon)}\sum\frac{bt^3}{3} \tag{5.2}$$

$$EC_w = E\frac{t_f b_f^3 (h_w + t_f)^2}{24} \tag{5.3}$$

where E is the Young's modulus, G is the shear modulus, υ is the Poisson's ratio, b_f is the flange width, t_f is flange thickness, and h_w is the web height of the IPG.

5.2   Lateral-torsional buckling of corrugated web girders

For the case of CWGs, Sause et al. [6] recommended to use the existing LTB equations of IPGs without considering the increased torsional stiffness of corrugated web girders until research demonstrates that CWGs have an increased inelastic LTB resistance compared to IPGs, although Lindner [5] found in 1990 this relative increase in CWGs LTB strengths. More recently, it has been found that CWGs, indeed, possess

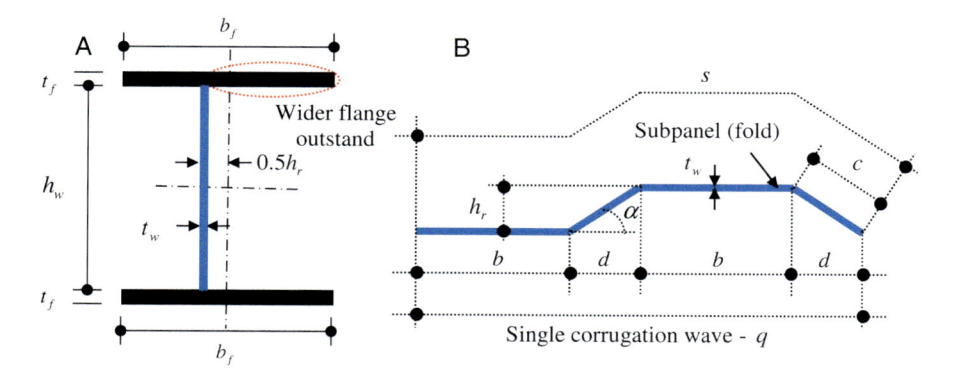

Figure 5.2 Steel I-section with corrugated web: (A) cross-section and (B) corrugation configuration and geometric notation.

increased LTB resistance compared to IPGs [8,9,11]. This has been attributed merely to the increase in the warping constant of the cross-section [8–9]. As a result, novel expressions for the warping constant ($C_{w,sc}$) have been proposed in the literature for CWGs, as given by (Eqs. 5.4–5.6), by Moon et al. [8] ($C_{w,mc}$), Nguyen et al. [9] ($C_{w,nc}$) and Lindner [5] ($C_{w,lc}$), respectively.

$$C_{w,mc} = \frac{I_{y,co}h_w^2}{4} + \frac{h_w^3 t_w}{12} \cdot \left(\frac{2b+d}{2b+2d}\right)^2 \cdot h_r^2 \tag{5.4}$$

$$C_{w,nc} = \frac{h_w^2}{24} \frac{b_f t_f}{\left(6b_f t_f + h_w t_w\right)} \left(6b_f^3 t_f + b_f^2 h_w t_w + 12 h_w t_w d^2\right) \tag{5.5}$$

$$C_{w,lc} = \frac{I_{y,co}h_w^2}{4} + \frac{h_r^2 h_w^2}{8\beta(b+d)} \cdot \frac{L^2}{\pi^2 E} \text{ with } \beta = \frac{h_w}{2Gbt_w} + \frac{h_w^2(b+d)^3(I_{x,co}+I_{y,co})}{600b^2 E I_{x,co} I_{y,co}} \tag{5.6}$$

where $I_{y,co}, I_{x,co}, L, h_w, t_w, b_f, t_f$ are the second moment of inertia of CWG about the weak axis, the second moment of inertia of CWG about the major axis, laterally unsupported length of the girder, web depth, web thickness, flange width, and thickness, respectively. Nomenclatures b and d are the widths of the flat and inclined folds of the corrugated web, respectively, as shown in Fig. 5.2. E and G are the Young's and shear moduli, respectively. Note that the format of $C_{w,mc}$, as represented in (Eq. 5.6), was given by Ibrahim [22], based on the simplification of the original method of calculation [8].

As can be seen, the warping constant proposed by Lindner [5] increases quadratically with the length of the girder (L), which was also observed by Larsson and Persson [21]. Accordingly, this warping constant is ignored in this book because a sectional constant should depend only on the cross-sectional geometrical details, and not on the unbraced length of the girder. It is worth pointing out that the studies on the increased LTB strength of CWGs were all dealing with those girders composed of normal-strength steel (NSS) [5,8–9].

5.3   High-strength steels in bridge construction

High strength steels (HSSs), with a nominal yield strength (F_y) of at least 460 MPa, have become widely commercially available thanks to the development in steel material technologies. Recently, it has been recognized that utilizing HSSs in bridges [23] and buildings [24] provides various structural benefits, such as increasing spans (i.e., providing column-free spaces) without using substantially thick steel plates. This enables the use of smaller cross-section sizes for the supporting structural elements, resulting in significant cost savings. Additionally, compared to conventional NSSs, the application of HSS reduces total steel consumption and CO_2 emissions in the building industry [25,26]. According to the World Steel Association [27], the application of HSS instead of NSSs reduces the product life cycle emission by 156 million tons CO_2 equivalents. Accordingly, using HSS in construction becomes essential to preserve the environment by effectively reducing carbon dioxide emissions from the steel industry.

Based on the above, a lot of bridges around the world have been built using HSSs [28], especially steels with a yield strength of 460 MPa, with the example of the Ilverich Bridge in Germany. Additionally, the demonstration bridge with corrugated webs shown in Fig. 5.3, which was designed by Pennsylvania Department of Transportation (PennDOT) and opened for service in July of 2005 [29], was formed from HSSs. However, as can be noticed above, LTB study of CWGs built up from HSSs has rarely been undertaken and existing researches using HSSs [6.23] have not shown the increased strength compared to IPGs.

A recent study on LTB of IPGs built up with HSSs by Bradford and Liu [30] has found that the design models in international specifications do not correspond to the ultimate moments of the girders. Therefore, modifications have been applied to these design models, raising the importance of investigating the members formed from HSSs. More recently, Somodi and Kövesdi [31] investigated the flexure buckling of HSS members, from which the applicable column buckling curves for steel grades between S420 and S960 were upgraded, compared to NSSs, using curves with smaller imperfection factors (α_{LT}) [32]. This is because the obtained buckling resistances for these steel grades were always higher than the appropriate EN 1993-1-1 [32] column buckling curve. This is mainly attributed to the fact that the residual stress amplitudes are much smaller for HSS structures compared to their yield stresses than those of NSS members [33]. Based on the above information, investigating LTB behavior of CWGs built up from HSSs was important to substitute the lack in their strength and behavior, which was done by Elkawas et al. [34].

However, as the material strength increases [4,34–36], the cost of the material usually increases. Nevertheless, if the higher strength can be fully utilized, the relative material cost reduces [37]. For that reason, a very cost-effective solution is to utilize HSS "hybrid girders," which are defined as those girders with flanges of HSS and webs of lower steel grade. The construction of several bridges proved that hybrid girders are more economical than homogeneous girders [37]. This was also been confirmed by

Figure 5.3 HSS demonstration bridge with corrugated webs in Bradford County, USA [29]. *HSS*, high strength steel.

IABSE which reported that bridges built in the USA based on hybrid design concept were more cost-effective [38]. As an example, a bridge was erected in Mttådalen, Sweden in 1995, with a span of 20 m and width of 7.0 m by mixing S690 and S460 in the girder's cross-section. An alternative design was made by considering S355 all over the cross-section, but the hybrid design was proved to be the most cost-effective [37]. Until 2020, no examination has been found in the literature investigating the hybrid (or even the homogenous) CWGs formed from steel S690 under LTB, despite early studies [39] on the bending behavior of hybrid steel I-girders with flat webs dating back to the 1964. Additionally, some valuable investigations have recently been undertaken on hybrid *flat-webbed IPGs* [40–41] in China and USA. This has recently encouraged Shao et al. [42] to go through the strength and behavior of laterally-unrestrained S690 high-strength steel hybrid girders with corrugated webs.

5.4 Homogenous corrugated web girders built up from high strength steels

5.4.1 Description of virtual tests

A parametric study was performed by Elkawas et al. [34] based on validated finite element (FE) models by comparisons with the test results of Driver et al. [35], considering both linear and non-linear buckling analyses available in ABAQUS [43]. This was to study how different geometrical parameters affect LTB critical and ultimate moments of CWGs. All FE models were simply-supported, in both flexure and torsion, and loaded under constant BM. It should be noted that all the girders were simulated using non-rigid end posts [23] with the supports located at distances of $h_w/10$ from both

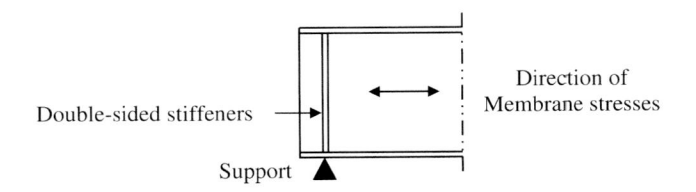

Figure 5.4 Considered end posts.

ends of the girders (Fig. 5.4). This is because the type of the end post depends on the longitudinal membrane stresses in the plane of the web. When these stresses are large, a rigid end post should be used, which adds double-sided stiffeners to the end of the girder. However, based on the negligible longitudinal stress induced in the corrugated webs, non-rigid end posts become the ideal choice [44].

During this parametric study, a single corrugated web taken similar to that practically used for Maupre bridge girders [45] was used. The corrugation dimensions of this bridge are $b = 284$ mm, $a = 241$ mm, $h_r = 150$ mm, $\alpha = 31.9°$ (see Fig. 5.2 for the geometric notations), with web height and thickness, respectively, of 2650 and 12 mm. The flange width and thickness were taken as 500 and 50 mm, respectively. Then, the parametric study was extended by changing only one parameter at a time as shown in Table 5.1 (given in bold). Currently, the parameter value ranges are:

- The girder length (L) is varying from 8520 to 21000 mm by increasing two corrugation waves in each model (i.e. from $L = 10q$ to $L = 20q$).
- The flange width (b_f) is changing from 300 to 500 mm with an increment of 100 mm.
- The flange thickness (t_f) has three values of 30, 40 and 50 mm.
- The web height (h_w) is changing from 1500 to 2650 mm.
- The web thickness (t_w) varies from 8 to 12 mm with increment of 2 mm,
- Three corrugation angles (α) are considered ranging from 31.9 to 60°, and
- The dimension (c) of the inclined fold as well as its projection in the longitudinal dimension (a) also varies as a result of changing the angle (α).

Table 5.2 provides the FE critical and ultimate BMs. The critical moments ($M_{cr,M}$) were calculated by the method of Moon et al. [8], using Eq. 5.1, and the nominal strengths based on the design model given by EC3 [32] for IPGs are also tabulated. The plastic strength of each cross-section ($M_{pl,R}$), which neglects the contribution from the web due to the accordion effect, is shown in Table 5.1.

5.4.2 Behavior of homogenous corrugated web girders built up with high strength steels

The critical buckling moments from the linear analyses were compared to the prediction of Moon et al. [8]. Overall, the suggested formula by Moon et al. [8] ($M_{cr,Moon}$) predicts the critical buckling conservatively with an average ratio for $M_{cr,FE}/M_{cr,Moon}$ of 1.03. However, as can be seen from Table 5.2, the ratio between critical FE moments with those predicted by the proposed method of Moon et al. [8] ($M_{cr,FE}/M_{cr,Moon}$) increases slightly with the increase in the girder's length (L).

Table 5.1 Details of the analyzed finite element models with fixed $b = 284$ mm and $c = 284$ mm.

Group	No.	α [°]	h_r [mm]	t_w [mm]	h_w [mm]	b_f [mm]	t_f [mm]	$M_{pl,R}$ [kN·m]	L [mm]
Basic group	1	31.9	150	12	2650	500	50	30,475	10,503
	2								12,604
	3								14,704
	4								16,805
	5								18,905
	6								21,006
Effect of b_f	7	31.9	150	12	2650	400	50	24,380	10,503
	8								12,604
	9								14,704
	10								16,805
	11								18,905
	12								21,006
	13	31.9	150	12	2650	300	50	18,285	10,503
	14								12,604
	15								14,704
	16								16,805
	17								18,905
	18								21,006

(*continued on next page*)

Table 5.1 Details of the analyzed finite element models with fixed $b = 284$ mm and $c = 284$ mm—cont'd

Group	No.	α [°]	h_r [mm]	t_w [mm]	h_w [mm]	b_r [mm]	t_f [mm]	$M_{pl,R}$ [kN·m]	L [mm]
Effect of h_w	19	31.9	150	12	2000	500	50	23,000	10,503
	20								12,604
	21								14,704
	22								16,805
	23								18,905
	24								21,006
	25	31.9	150	12	1500	500	50	17,250	10,503
	26								12,604
	27								14,704
	28								16,805
	29								18,905
	30								21,006
Effect of t_w	31	31.9	150	10	2650	500	50	30,475	10,503
	32								12,604
	33								14,704
	34								16,805
	35								18,905
	36								21,006
	37	31.9	150	8	2650	500	50	30,475	10,503
	38								12,604
	39								14,704
	40								16,805
	41								18,905
	42								21,006

(continued on next page)

Effect of t_f	43	31.9	150	12	2650	500	40	24,380	10,503
	44								12,604
	45								14,704
	46								16,805
	47								18,905
	48								21,006
	49	31.9	150	12	2650	500	30	18,285	10,503
	50								12,604
	51								14,704
	52								16,805
	53								18,905
	54								21,006
Effect of corrugation angle	55	45	150	12	2650	500	50	30,475	9698
	56								11,638
	57								13,577
	58								15,517
	59								17,456
	60								19,396
	61	60	150	12	2650	500	50	30,475	8523
	62								10,227
	63								11,932
	64								13,636
	65								15,341
	66								17,045

Table 5.2 Finite element strengths and comparisons of the analyzed finite element models with fixed $b = 284$ mm and $c = 284$ mm.

Group	No.	M_{cr} [kN·m]			M_{ul}[kN·m]		
		$M_{cr,FE}$	$M_{cr,M}$ [8]	$\dfrac{M_{cr,,FE}}{M_{cr,M}}$	$M_{ul,FE}$	M_{EC3} [32]	$\dfrac{M_{ul,FE}}{M_{EC3}}$
Basic group	1	26,171	26,203	1.00	20,717	15,111	1.37
	2	18,644	18,527	1.01	16,220	12,076	1.34
	3	14,117	13,893	1.02	12,843	9775	1.31
	4	11,125	10,880	1.02	10,277	8057	1.28
	5	9082	8810	1.03	8487	6765	1.25
	6	7615	7324	1.04	7183	5779	1.24
Effect of b_f	7	13,810	13,921	0.99	12,150	9242	1.31
	8	99,58	9923	1.00	9174	7159	1.28
	9	7625	7507	1.02	7129	5706	1.25
	10	6091	5933	1.03	5759	4673	1.23
	11	5021	4848	1.04	4841	3919	1.24
	12	4257	4067	1.05	4137	3353	1.23
	13	6221	6360	0.98	5830	4733	1.23
	14	4597	4596	1.00	4485	3604	1.24
	15	3584	3525	1.02	3444	2860	1.20
	16	2911	2825	1.03	2828	2347	1.20
	17	2442	2339	1.04	2398	1978	1.21
	18	2099	1988	1.06	2068	1704	1.21

(*continued on next page*)

Effect of h_w	19	20,582	20,306	1.01	16,280	11,582	1.41
	20	14,875	14,521	1.02	12,925	9356	1.38
	21	11,428	11,023	1.04	10,261	7664	1.34
	22	9141	8741	1.05	8481	6395	1.33
	23	7569	7168	1.06	7175	5437	1.32
	24	6433	6033	1.07	6165	4702	1.31
	25	16,494	15,972	1.03	12,793	8927	1.43
	26	12,163	11,618	1.05	10,463	7332	1.43
	27	9526	8971	1.06	8646	6110	1.42
	28	7759	7234	1.07	7209	5186	1.39
	29	6531	6027	1.08	6161	4481	1.37
	30	5633	5148	1.09	5425	3933	1.38
Effect of t_w	31	26,112	26,121	1.00	20,648	15,083	1.37
	32	18,594	18,466	1.01	16,220	12,049	1.35
	33	14,072	13,845	1.02	12,806	9749	1.31
	34	11,085	10,841	1.02	10,243	8033	1.28
	35	9045	8776	1.03	8455	6743	1.25
	36	7580	7294	1.04	7153	5759	1.24
	37	26,055	26,044	1.00	20,581	15,056	1.37
	38	18,546	18,410	1.01	16,113	12,023	1.34
	39	14,029	13,802	1.02	12,770	9726	1.31
	40	11,046	10,806	1.02	10,211	8012	1.27
	41	9009	8747	1.03	8425	6725	1.25
	42	7547	7270	1.04	7123	5742	1.24

(*continued on next page*)

Table 5.2 Finite element strengths and comparisons of the analyzed finite element models with fixed $b = 284$ mm and $c = 284$ mm—cont'd

| Group | No. | M_{cr} [kN·m] | | | M_{ul}[kN·m] | | |
		$M_{cr,FE}$	$M_{cr,M}$ [8]	$\dfrac{M_{cr,,FE}}{M_{cr,M}}$	$M_{ul,FE}$	M_{EC3} [32]	$\dfrac{M_{ul,FE}}{M_{EC3}}$
Effect of t_f	43	20,610	20,745	0.99	16,516	13,282	1.24
	44	14,584	14,583	1.00	12,801	10,490	1.22
	45	10,963	10,866	1.01	9855	8363	1.18
	46	8574	8452	1.01	7946	6790	1.17
	47	6945	6794	1.02	6568	5621	1.17
	48	5779	5607	1.03	5530	4740	1.17
	49	15,267	15,489	0.99	12,254	9935	1.23
	50	10,744	10,838	0.99	9314	7816	1.19
	51	8026	8034	1.00	7321	6200	1.18
	52	6234	6213	1.00	5822	5006	1.16
	53	5013	4964	1.01	4735	4120	1.15
	54	4140	4070	1.02	3947	3452	1.14
Effect of corrugation angle	55	30,821	30,876	1.00	22,963	16,570	1.39
	56	22,010	21,746	1.01	18,535	13,457	1.38
	57	16,669	16,237	1.03	14,860	10,988	1.35
	58	13,173	12,658	1.04	11,848	9093	1.30
	59	10,775	10,200	1.06	9985	7643	1.31
	60	9040	8438	1.07	8460	6523	1.30
	61	40,068	40,304	0.99	25,530	18,853	1.35
	62	28,647	28,255	1.01	22,106	15,783	1.40
	63	21,741	20,987	1.04	18,386	13,146	1.40
	64	17,199	16,268	1.06	15,251	11,004	1.39
	65	14,081	13,030	1.08	12,845	9302	1.38
	66	11,832	10,712	1.10	10,636	7955	1.34
Average				1.03			1.29
Standard deviation				0.027			0.078

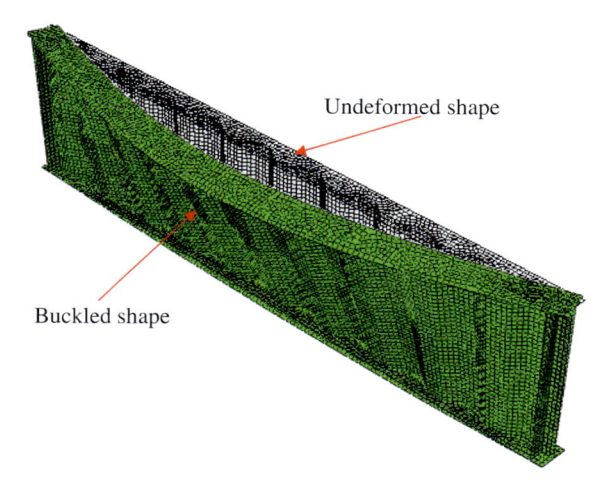

Figure 5.5 Typical LTB failure mode of CWGs. *CWGs*, corrugated web girders; *LTB*, lateral-torsional buckling.

It was found that CWGs buckle out-of-plan under the applied pure BMs, which resulted in LTB mode of failure. Fig. 5.5 shows the typical failure mode of current girders. As can be seen, the compression flange tends to destabilize when the length between the lateral supports exceeds the limit that produces flexural plastic hinge (the plastic length limit (L_p) in AISC [46] for example). The results showed that the buckling modes have shapes with maximum unit displacement amplitude and no vertical deflection, while the failure modes possess both the buckling mode seed and major-axis vertical deflection.

To explore the stress carried by the corrugated webs, Fig. 5.6 is added to show the stress contour of a typical girder (girder #1 in Table 5.1) captured at the ultimate moment ($M_{ul,FE}$). As can be seen, the web carries almost no BM except in regions very close to the flanges. The insignificant contribution of the web to the flexural strength is related to the accordion effect of the corrugated web. This means that the moment is completely carried by the upper and lower flanges in girders formed from HSSs. This is similar to the results available on LTB of corrugated web girders formed from normal mild steels [8].

With the aim of illustrating the general behavior of CWGs built up from HSSs, BM-in-plane-deflection relationships for the basic group are plotted in Fig. 5.7. It can be seen that the length of CWGs plays an important role in their behavior and strength. For shorter spans, the BM-in-plane-deflection curves drop sharply after reaching the ultimate strength ($M_{ul,FE}$) (see curves with $L = 10500$ mm to 14700 mm). In the case of larger spans, the ultimate strength ($M_{ul,FE}$) decreases steadily with the length increase as in CWGs with lengths ranging from 16800 to 21000 mm. These different responses are related to the type of LTB failure mode, which might take place inelastically (i.e. a certain amount of yielding will occur before ultimate BM) or elastically. The FE stress contour indicates that when the girder is slender enough to fail by the elastic

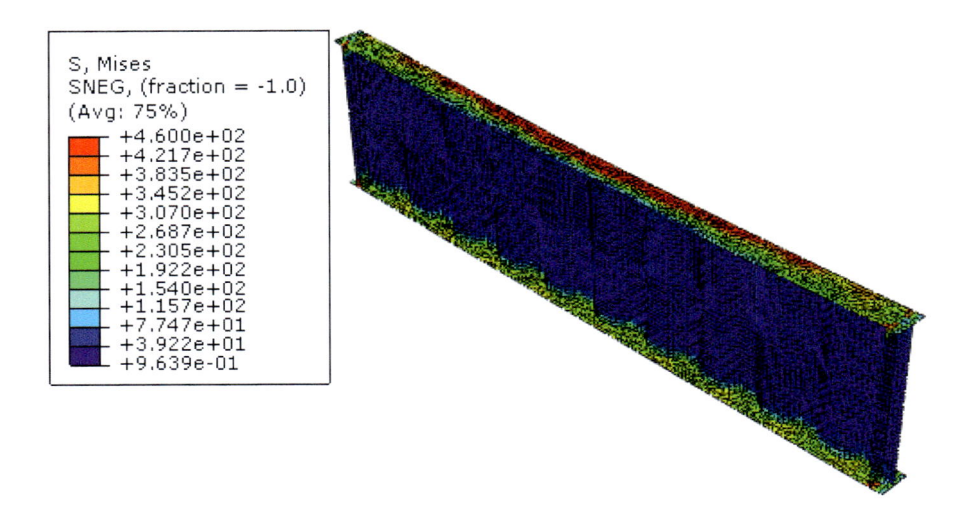

Figure 5.6 Typical deformed shape of CWGs [in MPa]. *CWGs*, corrugated web girders.

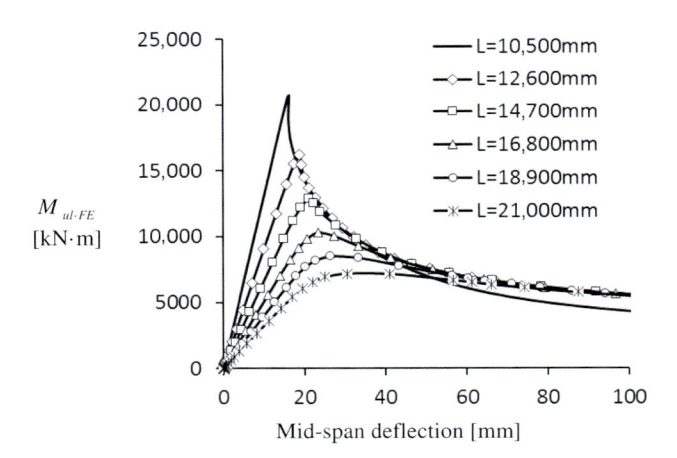

Figure 5.7 Relationships between LTB moment and the mid-span vertical deflection for the girders of the basic group. *LTB*, lateral-torsional buckling.

LTB mode, the failure becomes gradual because of the propagation of the second-order torques, due to the axial compression force in the compression flange and twist, from the beginning of the loading [17]. This, therefore, highlights the importance of arriving at the limit from which LTB occurs elastically, which is obtained at the end of this section. Fig. 5.7, additionally, shows that as girder length increases, the vertical deflection corresponding to the ultimate BM ($M_{ul,FE}$) becomes relatively larger. The stiffness of the curves is also influenced by the span length of the girders, from which it can be seen that it increases with length decrease.

5.4.3 Effects of key parameters

This subsection investigates the effects of the various key parameters on the behavior and strength of CWGs under pure BMs. In addition to the girder length (L) discussed in the previous subsection, the effects of the radius of gyration of the compression flange (r_T) and the section modulus (W_y) are indirectly evaluated. The effect of r_T is evaluated by changing the geometrical dimensions of the flange (t_f and b_f), while W_y is examined by changing the dimensions of the web plate (t_w and h_w). Additionally, the effect of the corrugation angle ($\alpha°$) is discussed here in detail. In the following subsections, each parameter is examined individually by keeping other variables constant.

5.4.3.1 Effect of flange geometrical dimensions

In this subsection, the effects of changing the flange width (b_f) and thickness (t_f) are presented. By varying the flange dimension, the flange slenderness ($b_f/2t_f$) ratios are evaluated accordingly. However, $b_f/2t_f$ ratios are believed to affect the local buckling behavior of the flanges, which is not the current case of evaluating LTB behavior of the girders. Fig. 5.8 displays $M_{ul,FE}$ values as well as $M_{ul,FE}/M_{pl,R}$ ratios against the length of the girders, considering different b_f and t_f values separately. The upper relationships consider the variation of b_f values, while the lower relationships show the effect of changing t_f values. From this figure, it can be noticed that $M_{ul,FE}$ values increase with the increase in the flange dimensions (b_f or t_f). However, the variation in $M_{ul,FE}$ values may be recognized to be relatively larger in the case of increasing the flange width (b_f) than in the case of increasing the thickness (t_f). This may be attributed to the effect of both parameters on the minor-axis second moment of area of the cross-section ($I_y = t_f b_f^3/6$), which in turn affects the value of the radius of gyration of the compression flange ($r_T = \sqrt{I_y/2A_f}$); A_f is the compression flange cross-sectional area. The computed value of I_y for the girders of the first group is 1042×10^6 mm^4. By changing the flange width to 400 and 300 mm, the I_y values become 533×10^6 and 225×10^6, respectively. On the other hand, the change in the flange thickness (from 50 mm) to 40 and 30 mm, respectively, provides I_y values of 833×10^6 and 625×10^6 mm^4. Accordingly, it can be recognized that the effect of b_f on I_y value is much greater than that of t_f. This indicates that the flexural capacity increases with increasing the value of r_T (i.e. the cross-section becomes stiffer in the torsion) which is consistent with the design procedure provided by AISC [46]. From the relative values to the left of the figure, it can clearly be noticed that CWGs with higher I_y values can attain higher flexural strengths compared to their $M_{pl,R}$ values. This provides weight and cost savings. On the other hand, this subsection also indicates the insignificant effect of $b_f/2t_f$ ratios as long as the girders are failing by LTB instead of local buckling.

Fig. 5.9 shows the relationship between the critical and the ultimate FE BMs against the flange width (b_f) of sample results. Plate girders with 10 corrugation waves in length (i.e. $L = 10q$), $t_w = 12$ mm, $h_w = 2650$ mm, $t_f = 50$ mm and with Maupre bridge corrugation profile ($b = 284$ mm, $a = 241$ mm, $\alpha = 31.6°$, $h_r = 150$ mm) are used in this figure. It can be recognized from this figure that both critical and ultimate BMs strongly depend on flange width. This confirms the results presented in the previous paragraph. Fig. 5.9 shows that increasing the flange width consequently increases both the critical and ultimate LTB moments. Additionally, it is noticeable that the critical and

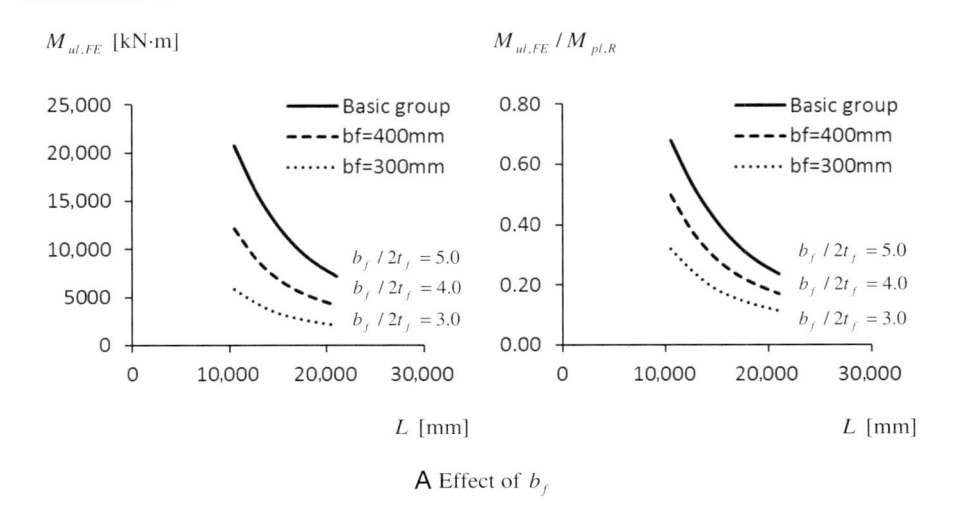

A Effect of b_f

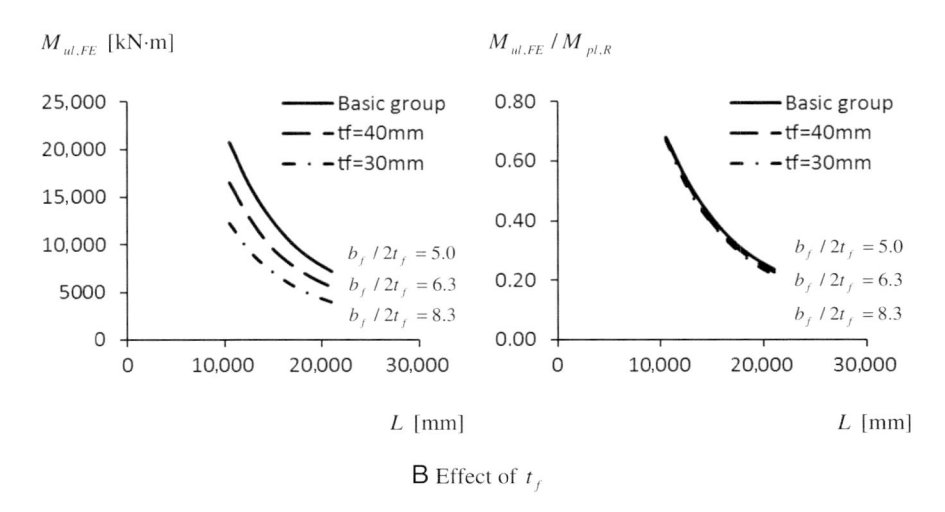

B Effect of t_f

Figure 5.8 Bending moment vs. unbraced length: (A) effect of b_f and (B) t_f.

ultimate LTB moments are very close to each other in the case of small flange widths. This may be attributed to the low slenderness ratio of the girder which reduces the applicability of the girder to make effective use of its material strength. Accordingly, a slenderness limit by which the use of the HSS material in CWGs becomes insignificant with respect to the strength enhancements is suggested in Section 5.6 of this chapter.

5.4.3.2　Effect of web plate dimensions (t_w and h_w)

The second important factor (beside the r_T discussed in the previous subsection) affecting the behavior and strength of CWGs is the plastic section modulus ($W_{pl,y}$). Therefore,

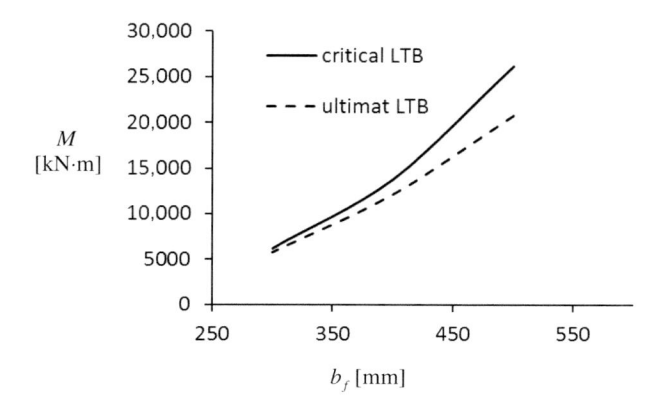

Figure 5.9 Critical/ultimate bending moments vs. flange width (b_f).

it is used in design according to EC3 [32], as will be noted later in this chapter. This modulus for flat-webbed girders is computed by considering the contributions of both the flanges and the web, with t_w and h_w both contributing effectively. For the case of CWGs, the effect of the web dimensions is considered here. It should be noted that changing both t_w and h_w affects the web slenderness (h_w/t_w) ratios, which (unlike $W_{pl,y}$) is not signifcant on the flexural buckling behavior of the girders. Alternatively, the web h_w/t_w ratios play important role in the shear buckling behavior of the current girders as can be realized from Elkawas et al. [36]. Accordingly, Fig. 5.10 is provided for this purpose by showing $M_{ul,FE}$ values as well as $M_{ul,FE}/M_{pl,R}$ ratios against the length of the girders, considering the different values of t_w and h_w individually. As can be seen, increasing h_w value has a positive effect on the strength of CWGs. However, the relative increase in $M_{ul,FE}$ relative to $M_{pl,R}$ (i.e., $M_{ul,FE}/M_{pl,R}$) shows that girders with shorter h_w utilize the material strength more effectively than those girders with higher web heights. On the other hand, the effect of the value of t_w is found to have no effect on the strength of CWGs. The results also show that the relationships between BM and mid-span vertical deflection for girders with different t_w values are identical. This is clearly attributed to the accordion effect. Hence, it can be concluded that a minimum value of t_w should be obtained in design by avoiding the overall shear buckling [3,4,45] of the webs without taking into account its effect on flexure.

5.4.3.3 Effect of corrugation angle ($\alpha°$)

The impact of the corrugation angle on LTB behavior of CWGs is discussed in this subsection. Three corrugation angles of $31.9°$, $45°$, and $60°$ were used. Bar charts are, therefore, presented for both ultimate ($M_{ul,FE}$) and normalized ($M_{ul,FE}/M_{pl,R}$) BMs, as can be seen in Fig. 5.11. From the figure, it can be recognized that increasing the corrugation angle definitely increases both $M_{ul,FE}$ values and $M_{ul,FE}/M_{pl,R}$ ratios. This may be attributed to the increase in the out-of-plane stiffness of the corrugated webs. Accordingly, increasing the corrugation angle is recommended to increase the flexural capacities of CWGs. Additionally, the unbraced length of the girder can be increased by using higher $\alpha°$ values without sacrificing the strength. For example, CWG with the length

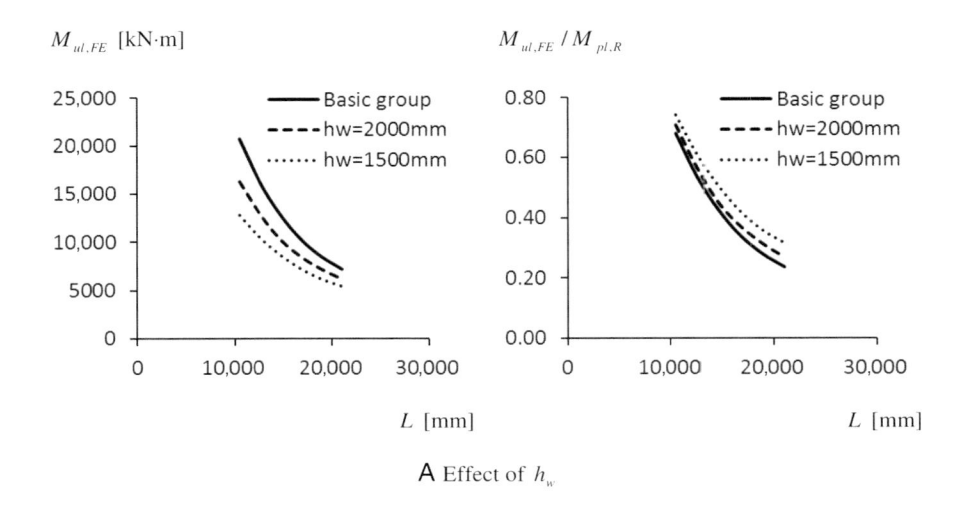

A Effect of h_w

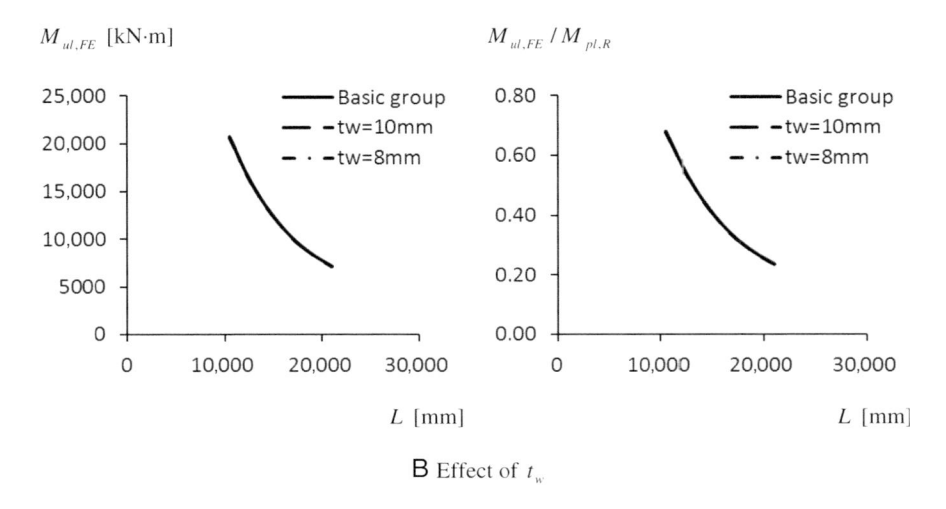

B Effect of t_w

Figure 5.10 Ultimate/normalized bending moments against the unbraced length: (A) effect of h_w and (B) t_w.

of 12 corrugation waves formed with corrugation angles of 60° has higher $M_{ul,FE}$ value than the shorter girder of 10 corrugated waves built with corrugation angles of 31.9°.

5.4.4 Comparison with EC3 design lateral-torsional buckling resistance

5.4.4.1 Design method

According to EN1993-1-1 [32], the unfactored design buckling resistance moment (denoted hereafter as $M_{EC3,Original}$) of a laterally unrestrained girder subjected to a

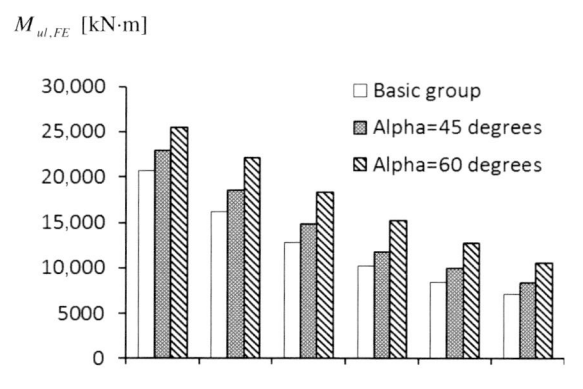

$M_{ul,FE}$ [kN·m]

L [mm]: 10 waves to 20 waves (with increment of 2 waves)

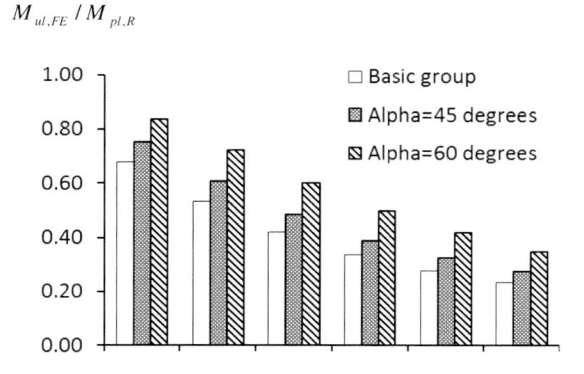

$M_{ul,FE}$ / $M_{pl,R}$

L [mm]: 10 waves to 20 waves (with increment of 2 waves)

Figure 5.11 Ultimate/normalized bending moment for different unbraced lengths showing the effect of $\alpha°$.

major-axis bending is given as:

$$M_{EC3} = \chi_{LT} W_y f_y \tag{5.7}$$

where:

$W_y = W_{pl,y}$ which is the plastic section modulus for Class 1 or 2 cross-sections, calculated herein by neglecting the contribution of the webs due to the accordion effect.

χ_{LT} is the reduction factor for LTB calculated from Clause 6.3.2.2 as:

$$\chi_{LT} = \frac{1}{\Phi_{LT} + \sqrt{\Phi_{LT}^2 - \overline{\lambda}_{LT}^2}} \leq 1.0 \tag{5.8}$$

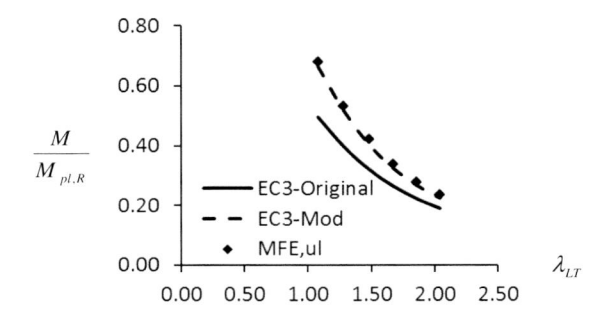

Figure 5.12 Comparison between normalized flexural strengths ($M_{ul,FE}$) and the original ($M_{EC3,Original}$) and modified ($M_{EC3,Mod}$) EC3 predictions—for the basic group with $b_f = 500$ mm, $h_w = 2650$ mm, $t_w = 12$ mm, $t_f = 50$ mm, $\alpha = 31.9°$.

where:

$$\Phi_{LT} = 0.5\left[1 + \alpha_{LT}\left(\overline{\lambda}_{LT} - 0.4\right) + \overline{\lambda}_{LT}^{2}\right] \qquad (5.9)$$

α_{LT} is an imperfection factor which has a value of 0.49 for welded open sections formed from steel S460 and $t_f > 40$ mm; buckling curve (c). For cases where $t_f \leq 40$ mm, buckling curve (b) was used with $\alpha_{LT} = 0.34$.

$$\overline{\lambda}_{LT} = \sqrt{\frac{W_y f_y}{M_{cr,LTB}}} \qquad (5.10)$$

$M_{cr,LTB}$ is the elastic critical moment for LTB which is given by (Eq. 5.1), with the warping constant calculated by using the formula suggested by Moon et al. [8] as described in Section 5.3.

5.4.4.2 Comparisons

Herein, the flexural strengths according to EN1993-1-1 [32] ($M_{EC3,Original}$) are normalized using the plastic strength of each cross-section ($M_{pl,R}$). It should be noted that the values of $M_{pl,R}$ and $M_{EC3,Original}$ for the considered groups are given in Table 5.2. The comparison of the numerical and predicted [32] normalized strengths is best observed with the help of Fig. 5.12 for the basic group, as other groups provide qualitatively similar results. The six data points, in each curve of Fig. 5.11, correspond to the FE models with the different considered lengths (see Table 5.1). It can easily be seen that the design equation given by EN 1993-1-1 [32] provides strengths ($M_{EC3,Original}$) on the safe side relative to FE results, but with very high conservative values. However, more accurate results are reached by increasing the unbraced lengths (L_b) of the girders. The average of $M_{ul,FE}/M_{EC3,Original}$ ratios was calculated and found to be 1.29 (ranging from 1.14 to 1.43) with a standard deviation of 0.08. Accordingly, a modified strength may better be suggested for the economical design of CWGs built up from HSSs. This is, simply, done here by applying the imperfection factor (α_{LT}) of

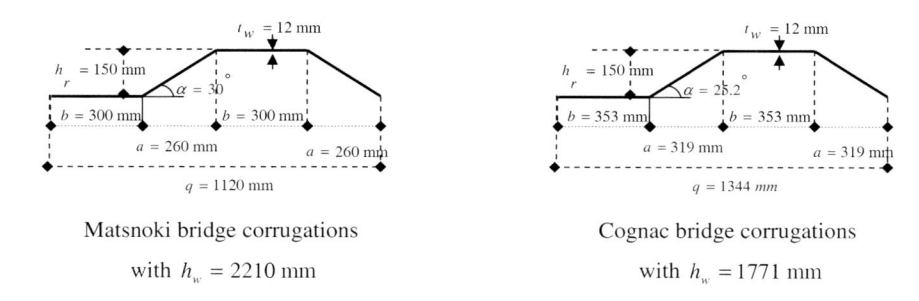

Figure 5.13 Corrugation dimensions used to find the elastic slenderness limit.

0.13 to the design equations provided by EC3 [32]. This value corresponds to buckling curve (a_o). This modified strength is also represented in Fig. 5.12, from which it can be seen that it provides excellent results with an average ratio of 1.05 for $M_{ul,FE}/M_{EC3,\text{Mod}}$ ratios (varying from 1.01 to 1.15) with a standard deviation of 0.03. However, until additional experiments become available for CWGs built up from HSSs, it is currently suggested to be more conservative and to use buckling curve (a) instead of (a_o) used in the above comparisons. This is consistent with the results of Somodi and Kövesdi [31] who found that the obtained flexural buckling resistance for steel grades up to S700 is always higher than EC3 [32] column buckling curve (b).

5.4.5 Slenderness limit of inelastic lateral-torsional buckling

Defining the elastic LTB limit is important because beyond this limit M_{cr} value becomes physically the failure moment, from which the load-deflection relationships become different from those girders failing inelastically as shown previously in Subsection 5.4.1. So, this elastic limit is examined by generating some new models considering the three corrugation dimensions belonging to Maupre, Matsnoki, and Cognac bridges [3,45], from which the slenderness parameter varied between 1.0 and 3.0. Fig. 5.13 provides the corrugation dimensions of Matsnoki and Cognac bridges, as those of Maupre corrugations were given previously in Section 5.3. Herein, the critical-to-ultimate FE strength ($M_{cr,FE}/M_{ul,FE}$) ratios are plotted in Fig. 5.14 against the slenderness parameter ($\overline{\lambda}_{LT}$) calculated by Eq. (5.7). A polynomial trend line representing the current relationship is also drawn in the figure for the results of each group. It can be seen that CWGs behave elastically, providing the condition of $M_{cr,FE}/M_{ul,FE} \leq 1.0$, when $\overline{\lambda}_{LT} \geq 2.2$. After this limit ($\overline{\lambda}_{LT,el} = 2.2$), using HSS material in CWGs becomes ineffective. Accordingly, it is recommended not to use CWGs built up from HSS with $\overline{\lambda}_{LT} \geq 2.2$ provided that the failure mode is obtained by LTB.

5.5 Hybrid corrugated web girders built up from high strength steels

5.5.1 Effect of corrugated web material

A study into the effect of web material grade on the strength and behavior of CWGs composed of S690 flanges was performed by Shao et al. [42]. To do so, FE models

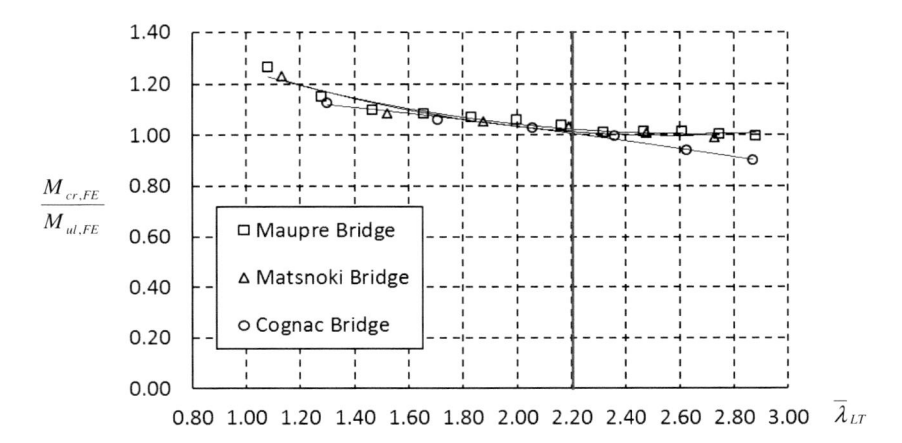

Figure 5.14 Critical-to-FE strength vs. slenderness parameter for determining the elastic slenderness limit. *FE*, finite element.

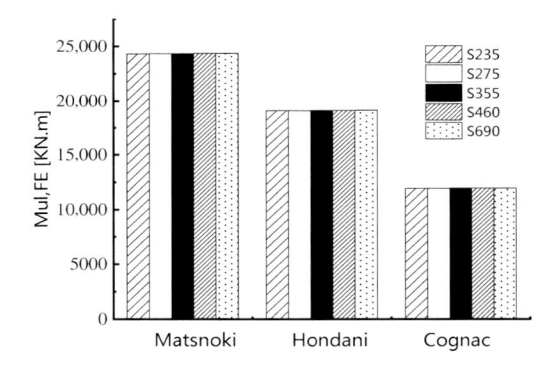

Figure 5.15 Effect of web material grade on the ultimate moment of CWGs. *CWGs, corrugated web girders.*

were generated considering the corrugation dimensions of three bridge dimensions; Matsnoki, Hondani, and Cognac bridges. Flanges of 500 mm width and 50 mm thickness were considered. All girders were formed from 10 corrugation waves to ensure that all girders fail by inelastic LTB. For each corrugation configuration, the web material has been considered as S235, S275, S355, S460 [32], and S690 [47], while the flanges were retained as S690 in the whole girders. The results confirmed that the material grade of the corrugated web has no effect on the strength and behavior of CWGs. Fig. 5.15 shows $M_{ul,FE}$ values of the current girders, from which it is clear that changing the grade of the web material has not affected the girders' strengths. This means that the corrugated webs provide the cross-section with superior out-of-plane stiffness based on their innovative folding [8], while the yield stress of the steel material has no effect. Additionally, for each group of models, the moment-deflection responses have coincided. This means that the ductility of the girders is not affected by the hybrid ratio, which is the ratio between the yield stress of the flange to that of the web, in contrast to the case of the flat-webbed hybrid girders [41]. Hence, hybrid CWGs (having the same flange steel grade) operate identically under earthquake conditions. Therefore, it would be excellent to use webs of steel S235 to form the hybrid section with flanges from S690, provided that LTB dominates the failure. By doing so, the material cost of the webs will be reduced on average by 37% compared to using steel S690, according to the Chinese steel market. Despite that, the limitations of Swedish codes [37], which define the maximum difference in steel grades, are followed in this book until experimental results in future confirm the negligible effect of the hybrid ratio. According to Swedish codes [37], since it is common practice in Sweden to use hybrid girders, it is required that the web's strength be not less than 50% of the flange's strength, due to serviceability issues [37]. Additionally, the shear strength of the webs should not be reduced too much by using the least steel grade [4,35], even if there is no shear acting currently, based on the positive effect of raising the web's steel grade to sustain the shear effectively. Therefore, for the case of current hybrid girders with flanges formed from S690, in the next sections, steel grade of S355 for the webs was used. In this case, the average cost of webs formed with a yield stress of 345 MPa is reduced by 30% compared to those with a yield stress of 690 MPa. Accordingly, this combination of different steel grades provides cost-effective cross-sections for CWGs, without sacrificing their strengths, and at the same time it conforms to some international codes [37].

5.5.2 Behavior of hybrid corrugated web girders built up with high strength steels

The fundamental behavior of the hybrid CWGs is provided, considering girders with flanges and webs made from S690 and S355, respectively. The fundamental behavior was investigated mainly through changing the slenderness of the girder ($\overline{\lambda}_{LT} = \sqrt{W_y F_y / M_{cr,LTB}}$) [32]; where W_y is the section modulus and $W_y F_y$ represents the plastic BM (M_{pl}) of section. Note that the calculation of M_{pl}, unlike that of IPGs with flat webs [41], ignores the contribution of the web due to the accordion effect of CWGs [48,49]. $M_{cr,LTB}$ values are currently utilizing the warping constant suggested by Moon et al. [8]. Accordingly, five main groups of models (G1–G5), each representing the dimensions of a constructed bridge, as shown in Table 5.3, were simulated by Shao et al. [42]. The corrugation dimensions are represented in Fig. 5.16, with b_f and t_f were taken as 400 mm and 40 mm, respectively. Note that t_f of 40 mm was considered to eliminate the thickness effect of the steel material from the current FE models, so that the nominal yield stress of steel S690 as specified by EC3 [50] can be used. This is because the yield stress of steel plates with thicknesses larger than 50 mm should be reduced [50]. Note that the material properties of the short tests of Sun et al. [47] were used. Accordingly, the values of M_{pl} are 24,844 MPa, 37,266 MPa, 19,909 MPa, 28,621 MPa, and 25,766 MPa for the cross-sections of Matsnoki, Hondani, Cognac, Dole, and IIsun bridges, respectively. From the table, it can be seen that each group consists of nine FE models by vaying the number of waves. Note that the shortest girder in each group was taken

Table 5.3 Details of the designed finite element models for the fundamental behavior of the corrugated web girders considering the nonlinear analysis.

Bridges	No.	waves	L [mm]	$M_{cr,FE}$ [kN·m]	$M_{cr,M}$ [kN·m]	$M_{ul,FE}$ [kN·m]	$M_{n,EC3}$ [kN·m]	$\bar{\lambda}_{LT}$	$\dfrac{M_{cr,FE}}{M_{cr,M}}$	$\dfrac{M_{cr,FE}}{M_{ul,FE}}$	$\dfrac{M_{ul,FE}}{M_{n,EC3}}$
G1: Matsnoki	1	4	4480	50,409	52,041	23,106	16,126	0.69	0.97	2.18	1.43
	2	6	6720	23,058	23,440	18,119	11,237	1.03	0.98	1.27	1.61
	3	8	8960	13,368	13,435	11,769	7906	1.36	1.00	1.14	1.49
	4	10	10,500	8849	8799	7998	5791	1.68	1.01	1.11	1.38
	5	12	12,600	6384	6277	5907	4426	1.99	1.02	1.08	1.33
	6	14	14,700	4885	4752	4871	3512	2.29	1.03	1.00	1.39
	7	16	16,800	3906	3759	4278	2873	2.57	1.04	0.91	1.49
	8	18	18,900	3227	3074	4147	2411	2.84	1.05	0.78	1.72
	9	20	21,000	2734	2581	3908	2065	3.10	1.06	0.70	1.89
G2: Hondani	10	4	4800	65,756	70,114	33,921	23,266	0.73	0.94	1.94	1.46
	11	6	7200	29,816	31,348	24,543	15,776	1.09	0.95	1.21	1.56
	12	8	9600	17,095	17,792	15,144	10,858	1.45	0.96	1.13	1.39
	13	10	12,000	11,175	11,516	10,020	7807	1.80	0.97	1.12	1.28
	14	12	14,400	7959	8105	7171	5867	2.14	0.98	1.11	1.22
	15	14	16,800	6003	6047	5710	4578	2.48	0.99	1.05	1.25
	16	16	19,200	4733	4710	4839	3684	2.81	1.00	0.98	1.31
	17	18	21,600	3856	3793	4367	3041	3.13	1.02	0.88	1.44
G3: Cognac	19	4	5376	28,572	29,086	17,066	11,211	0.83	0.98	1.67	1.52
	20	6	8064	13,278	13,316	11,377	7310	1.22	1.00	1.17	1.56
	21	8	10,752	7877	7791	7102	5008	1.60	1.01	1.11	1.42
	22	10	13,440	5345	5223	5002	3657	1.95	1.02	1.07	1.37
	23	12	16,128	3955	3818	4230	2821	2.28	1.04	0.93	1.50
	24	14	18,816	3099	2963	4090	2270	2.59	1.05	0.76	1.80
	25	16	21,504	2534	2400	4042	1888	2.88	1.06	0.63	2.14
	26	18	24,192	2137	2008	3711	1612	3.15	1.06	0.58	2.30

(*continued on next page*)

Table 5.3 Details of the designed finite element models for the fundamental behavior of the corrugated web girders considering the nonlinear analysis—cont'd

Bridges	No.	waves	L [mm]	$M_{cr,FE}$ [kN·m]	$M_{cr,M}$ [kN·m]	$M_{ul,FE}$ [kN·m]	$M_{n,EC3}$ [kN·m]	$\bar{\lambda}_{LT}$	$\dfrac{M_{cr,FE}}{M_{cr,M}}$	$\dfrac{M_{cr,FE}}{M_{ul,FE}}$	$\dfrac{M_{ul,FE}}{M_{n,EC3}}$
G4: Dole	28	4	6400	29,259	30,652	21,862	13,869	0.97	0.95	1.34	1.58
	29	6	9600	13,482	13,884	11,681	8435	1.44	0.97	1.15	1.38
	30	8	12,800	7906	8016	6851	5544	1.89	0.99	1.15	1.24
	31	10	16,000	5313	5295	5016	3933	2.32	1.00	1.06	1.28
	32	12	19,200	3890	3813	4398	2963	2.74	1.02	0.88	1.48
	33	14	22,400	3018	2914	4146	2337	3.13	1.04	0.73	1.77
G5: IIsun	37	4	5280	38,789	41,191	22,704	15,081	0.79	0.94	1.71	1.51
	38	6	7920	17,937	18,647	15,117	9950	1.18	0.96	1.19	1.52
	39	8	10,560	10,543	10,758	9229	6804	1.55	0.98	1.14	1.36
	40	10	13,200	7079	7100	6276	4926	1.90	1.00	1.13	1.27
	41	12	15,840	5194	5107	5047	3753	2.25	1.02	1.03	1.34
	42	14	18,480	4028	3900	4524	2981	2.57	1.03	0.89	1.52
	43	16	21,120	3266	3111	4283	2447	2.88	1.05	0.76	1.75
	44	18	23,760	2733	2566	3845	2062	3.17	1.06	0.71	1.86
Average									1.00		1.52
Standard deviation									0.036		0.240

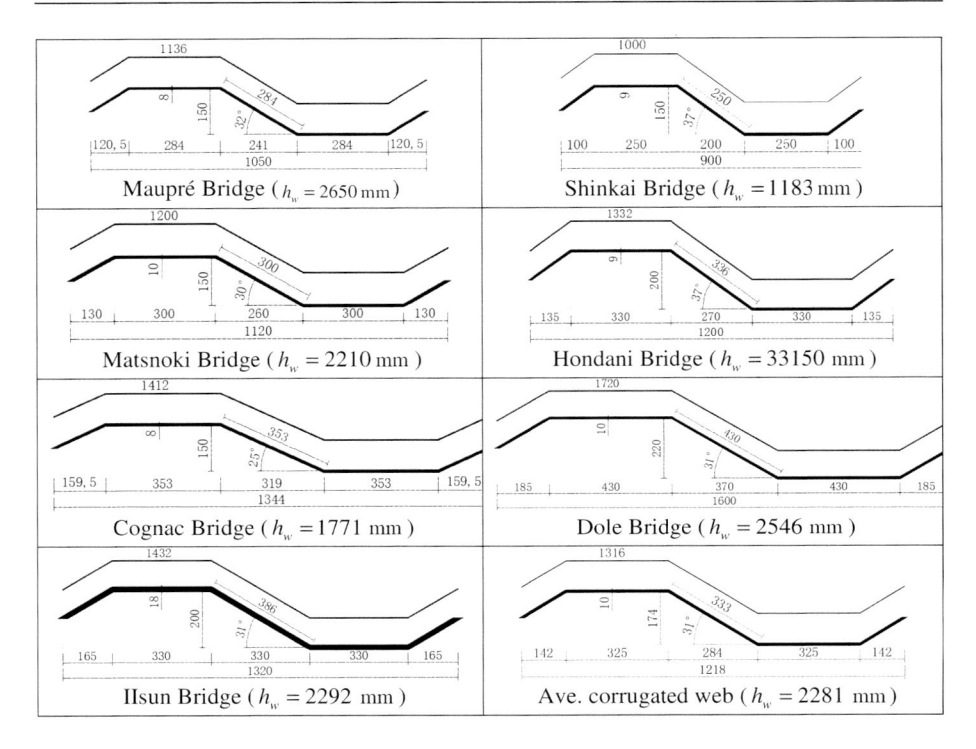

Figure 5.16 Details of corrugated webs used on the validation of the buckling analyses.

of four corrugation waves to eliminate the effects of the supports from the critical load amplifier, as suggested by Jáger et al. [51]. Then, the length of the girders increases by two waves several times, until the slenderness ($\overline{\lambda}_{LT}$) of the longest girder exceeds 3.0. This value guarantees that the elastic failure is reached in each group. According to the described dimensions, the slenderness of the girders ($\overline{\lambda}_{LT}$) had minimum and maximum values of 0.69 and 3.17, respectively. From Table 5.3, it can again be ensured that $M_{cr,M}$ values represent well the FE critical strengths ($M_{cr,FE}$), with an average ratio of 1.0 and a standard deviation of 0.036. Additionally, it can be seen that the girders of each group of models fail either inelastically (with $M_{cr,FE}/M_{ul,FE} > 1.0$) or elastically (with $M_{cr,FE}/M_{ul,FE} < 1.0$).

Examples of stress distributions for girders failing by inelastic and elastic LTB are shown in Fig. 5.17, considering Hondani Bridge girders #11 and #16. Both stress distributions were captured at the ultimate loads of the girders. From this figure, both girders as can be noticed failed by LTB. However, the short girder failed after some portions of the compression flange reached the yield stress, while the longer girder failed without reaching the yield stress. Additionally, the stress distributions for both girders confirm that the corrugated web does not withstand longitudinal stresses due to flexure except in very small areas near both flanges. The moment against the mid-span vertical displacement relationships for the whole group of Hondani Bridge is given in

A FE model #11: Hondani Bridge

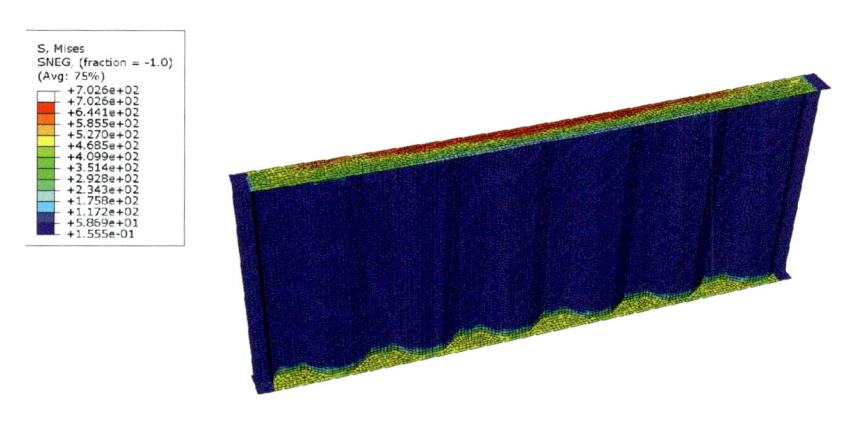

B FE model #16: Hondani Bridge

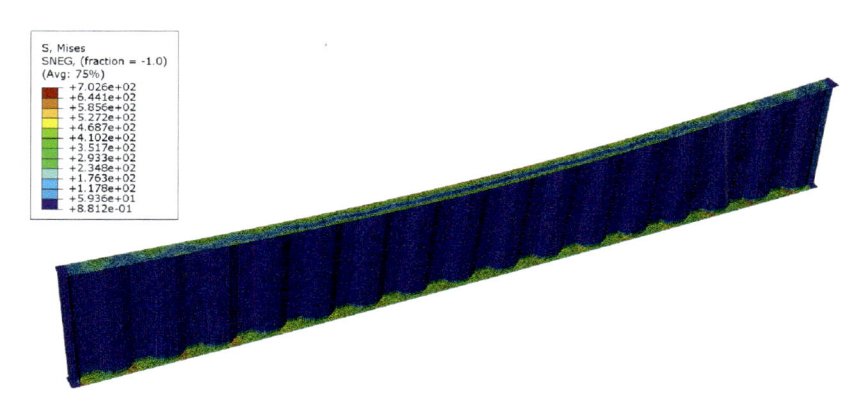

Figure 5.17 Failure modes observed for relatively short and long CWGs. *CWGs*, corrugated web girders.

Fig. 5.18. Based on this figure, it is evident that the shorter beam (in Fig. 5.17) fails dramatically after reaching the ultimate BM, while the longer girder (i.e., #16) failed in a ductile manner. This is because of the propagation of the second-order torques from the beginning of loading.

5.5.3 Effects of key parameters

This section provides some additional parametric studies demonstrating the effects of some variables on the strengths of the S690 hybrid CWGs. So, girder #4 in Table 5.3, which belongs to G1 with Matsnoki corrugation configuration, was currently chosen as a reference. Then, four additional groups (G6-G9) were generated to evaluate the

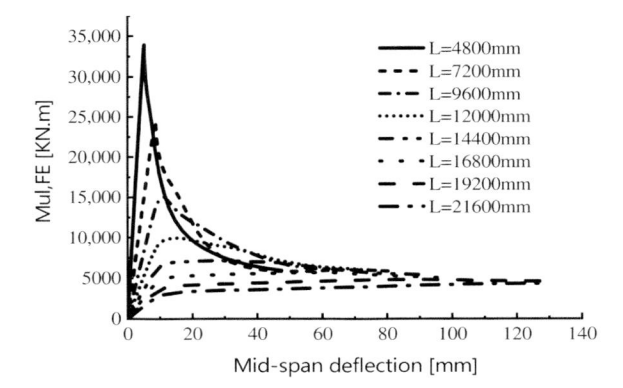

Figure 5.18 LTB moment vs. mid-span vertical deflection for the girders of Hondani group. *LTB*, lateral-torsional buckling.

effects of changing the corrugation depth (h_r), web thickness (t_w), and flange width (b_f). During the choice of the dimensions of the groups, the corrugation wavelength (q) was intentionally kept constant in all groups for comparison purposes and it was equal to 1120 mm. In group G6, the values of h_r increased (from 150 mm to 266 mm) with the corrugation angle (α) remaining constant at 30°. So, the corrugation dimensions *a,b,and c* (shown in Fig. 5.2B) were changed, as can be seen in Table 5.4. For the case of group G7 (Table 5.5), the change in h_r was associated with a change in the value of α (changed from 30° to 46.1°). Accordingly, dimensions *a* and *b* remained unchanged, while the width of the inclined fold (*c*) tended to change. As can be seen in Table 5.6, Group 8 was generated only by increasing the value t_w from 10 mm to 18 mm, while other dimensions were kept constant. Finally, the variation in group 9 (Table 5.7) was in the value of b_f, which has been increased from 400 mm to an upper limit of 600 mm which ensured that all the girders are formedfrom fully-effective flanges. In Tables 5.4–5.7, the values of $\bar{\lambda}_{LT}$ and $M_{ul,FE}$ are provided. Additionally, the % increase in the strength of each girder with respect to the ultimate strength of the reference girder was calculated and added to the tables. Furthermore, the contribution of the web in the warping constant over that of the flange ($C_{w,co,w}/C_{w,co,f}$) is added to the table. Additional two columns have been added to the tables to show the weight % of the web and flanges in each girder.

The results of Tables 5.4–5.6 indicate the negligible effect of changing any dimension related to the webs of the girders (i.e. h_r, t_w or α) on the strength of the girders or the weight percentage of the web and flanges. This is attributed to the fact that the effect of h_r, t_w, and α on the value of the warping constant is very small, as can be seen from the table, putting in mind that the larger capacities of CWGs compared to IPGs are related to the increase in the warping constant ($C_{w,co}$) [5]. As can be noticed, the maximum increase in $M_{ul,FE}$ value in groups G6–G8 is about 4.0% in G7, and it took place for the girder having the highest $C_{w,co,w}/C_{w,co,f}$ ratio. On the contrary, Table 5.7 confirms that increasing the flange width (b_f) of CWGs is an effective method in increasing the

Table 5.4 Effect of increasing h_r with α kept constant on the strengths and components' weights of the corrugated web girders—G6.

No.	Plan of one corrugation wave (containing the flange)	λ_{LT}	$M_{ul,FE}$ [kN·m]	M% increase relative to #1	$\dfrac{C_{w,co,w}}{C_{w,co,f}}$ %	Weight % Web	Weight % Flanges
1		1.68	7998	–	5.7	42.5	57.5
2		1.68	8052	0.7	7.2	42.8	57.2

(continued on next page)

Table 5.4 Effect of increasing h_r with α kept constant on the strengths and components' weights of the corrugated web girders—G6—cont'd

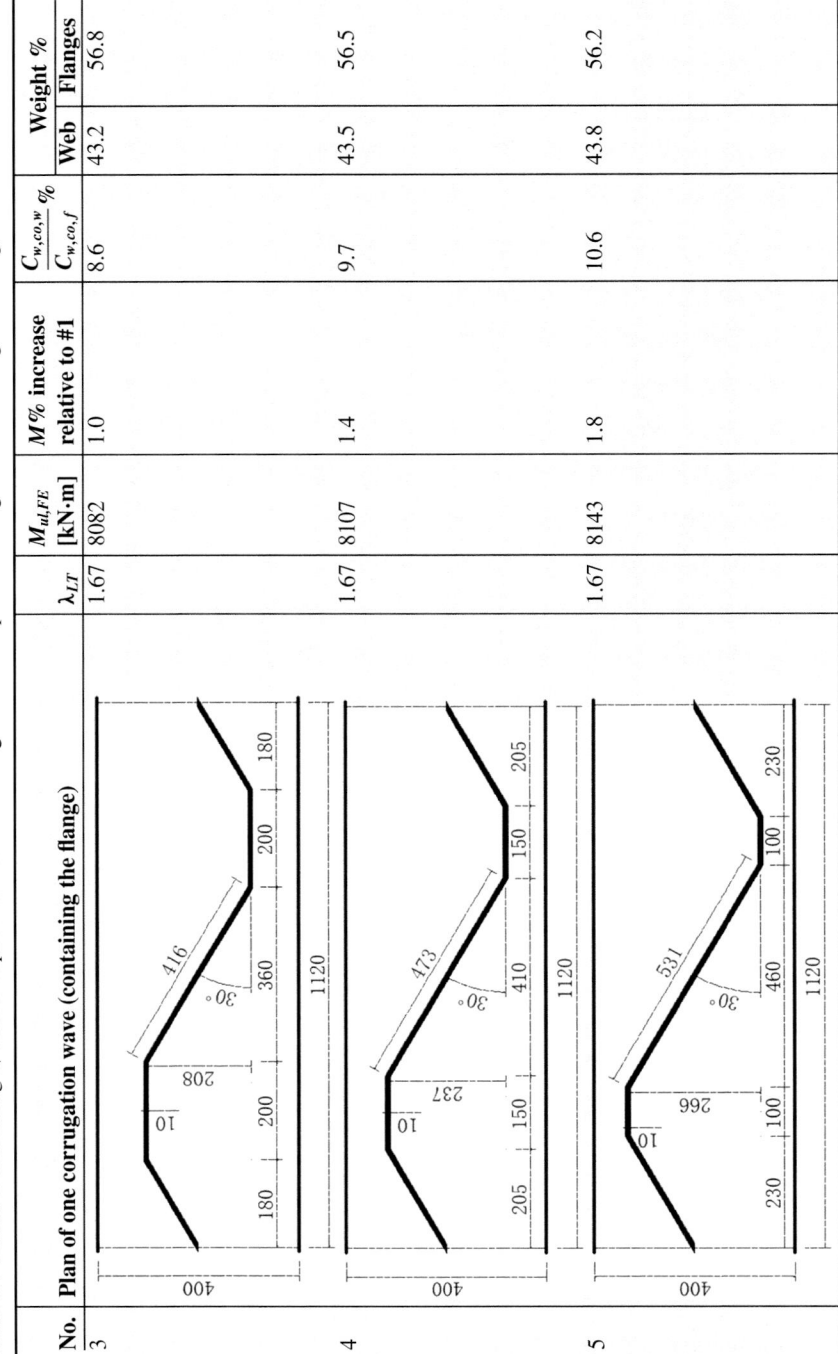

No.	Plan of one corrugation wave (containing the flange)	λ_{LT}	$M_{ul,FE}$ [kN·m]	M% increase relative to #1	$\dfrac{C_{w,co,w}}{C_{w,co,f}}$ %	Weight % Web	Weight % Flanges
3		1.67	8082	1.0	8.6	43.2	56.8
4		1.67	8107	1.4	9.7	43.5	56.5
5		1.67	8143	1.8	10.6	43.8	56.2

Table 5.5 Effect of increasing h_r with α changes on the strengths and components' weights of the corrugated web girders—G7.

No.	Plan of one corrugation wave (containing the flange)	λ_{LT}	$M_{ul,FE}$ [kN·m]	$M\%$ increase relative to #1	$\dfrac{C_{w,co,w}}{C_{w,co,f}}\%$	Weight %	
						Web	Flanges
6		1.67	8071	0.9	8.2	43.2	56.8
7		1.66	8134	1.7	11.2	43.9	56.1

(*continued on next page*)

Table 5.5 Effect of increasing h_r with α changes on the strengths and components' weights of the corrugated web girders—G7—cont'd

No.	Plan of one corrugation wave (containing the flange)	λ_{LT}	$M_{ul,FE}$ [kN·m]	M% increase relative to #1	$\dfrac{C_{w,co,w}}{C_{w,co,f}}$ %	Weight %	
						Web	Flanges
8		1.66	8221	2.8	14.7	44.6	55.4
9		1.64	8318	4.0	18.6	45.4	54.6

Table 5.6 Effect of increasing t_w on the strengths and components' weights of the corrugated web girders—G8.

No.	Thickness of corrugated web using the corrugation configuration of the reference girder	λ_{LT}	$M_{ul,FE}$ [kN·m]	M% increase relative to #1	$\dfrac{C_{w,co,w}}{C_{w,co,f}}$ %	Weight %	
						Web	Flanges
10	Web thickness is 12 mm	1.67	8055	0.7	6.9	47.0	53.0
11	Web thickness is 14 mm	1.67	8068	0.9	8.0	50.9	49.1
12	Web thickness is 16 mm	1.66	8103	1.3	9.2	54.2	45.8
13	Web thickness is 18 mm	1.65	8150	1.9	10.3	57.1	42.9

flexural strength of the girders. This is because the increase in the strength is much higher than the increase in the weight % of the flanges. Accordingly, by increasing the value of b_f in S690 hybrid CWGs, the material can be used effectively, and at the same time the environmental positive effect (by decreasing the emissions of the carbon dioxide as described earlier) of using the HSS increases.

5.5.4 Design strengths

The design strengths ($M_{n,EC3}$) calculated by EC3 [32], previously reported in Section 5.4.4.1, and $M_{ul,FE}/M_{n,EC3}$ ratios are given in Table 5.3. From this table, it is evident that the design method of EC3 [32] is extremely conservative compared to FE strengths (with an average $M_{ul,FE}/M_{n,EC3}$ ratio of 1.52 and a standard deviation of 0.240). This is due to the fact that this design method was proposed based on experimental tests on specimens formed from NSSs (see for example [52]). Hence, the suitability of current design provisions to S690 HSS requires additional revision.

For the case of the current CWGs, a design strength ($M_{n,Sug}$) is suggested, as given by Eq. (5.11). Herein, the comparative results between current ($M_{n,EC3}$) and suggested ($M_{n,Sug}$) design models are shown in Fig. 5.19. This equation uses two different design models for both inelastic and elastic LTB failure modes. For CWGs with inelastic LTB mode, it is suggested to use buckling curve (a) instead of buckling curve (d). So, the imperfection factor (α_{LT}) becomes 0.21. On the other hand, for CWGs failing with elastic LTB, it is suggested directly to use the critical buckling moment ($M_{cr,M}$) which uses the warping constant suggested by Moon et al. [8]. At this range $\overline{\lambda}_{LT} \geq 2.3$, using $M_{cr,M}$ values results in significantly better strengths compared to those calculated using the modified $M_{n,EC3}$ values with buckling curve (a). Overall, the average $M_{ul,FE}/M_{n,EC3}$ ratio is 1.17 with a standard deviation of 0.199. However, since the current S690 HSS CWGs with $\overline{\lambda}_{LT} \geq 2.3$ are not preferred to be used, the equation for CWGs failing elastically should be improved in the future.

$$M_{n,Sug} = \begin{cases} \chi_{LT}M_y \text{ (with buckling curve }'a')\ast: \lambda_{LT} < 2.3 \\ M_{cr,M} \qquad\qquad\qquad\qquad\quad : \lambda_{LT} \geq 2.3 \end{cases} \tag{5.11}$$

Table 5.7 Effect of increasing b_f on the strengths and components' weights of the corrugated web girders—G9.

No.	Plan of one corrugation wave (containing the flange)	λ_{LT}	$M_{ul,FE}$ [kN·m]	$M\%$ increase relative to #1	$\dfrac{C_{w,co,w}}{C_{w,co,f}}\%$	Weight %	
						Web	Flanges
14		1.51	11,153	39.5	4.0	39.7	60.3
15		1.37	14,895	86.2	2.9	37.2	62.8

(*continued on next page*)

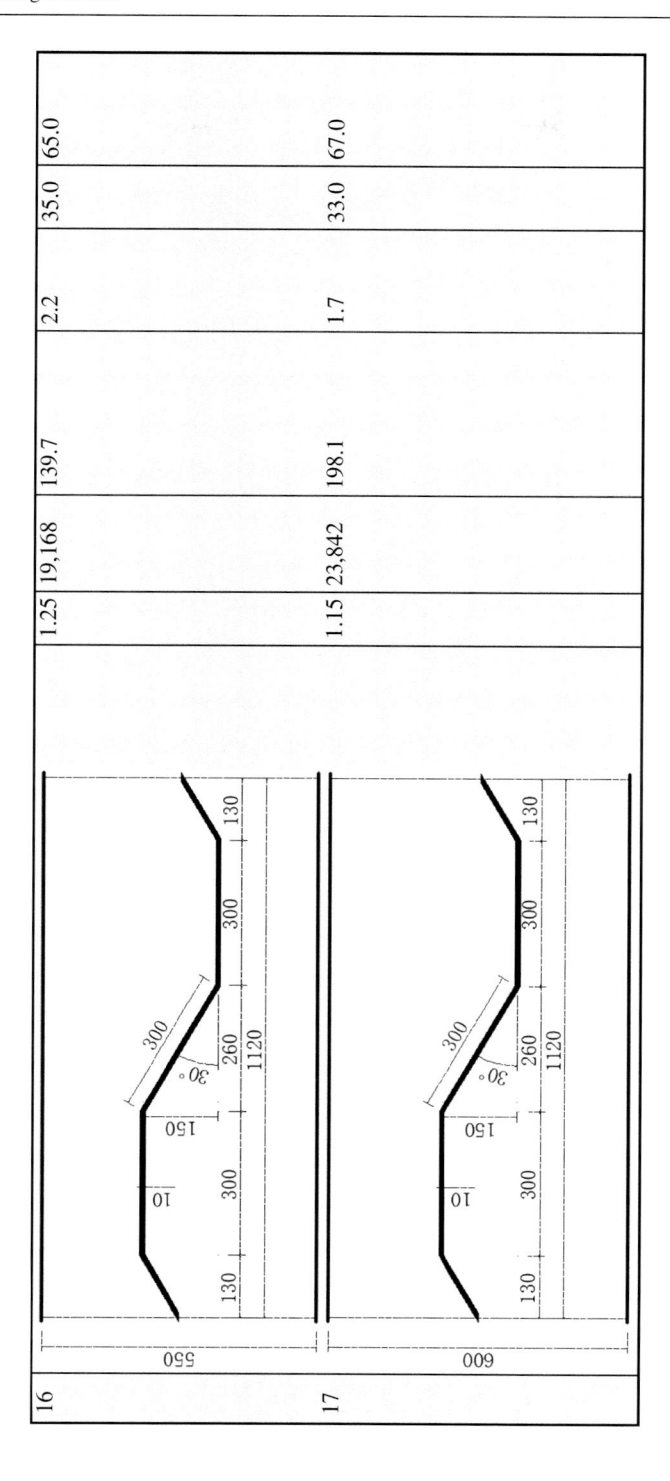

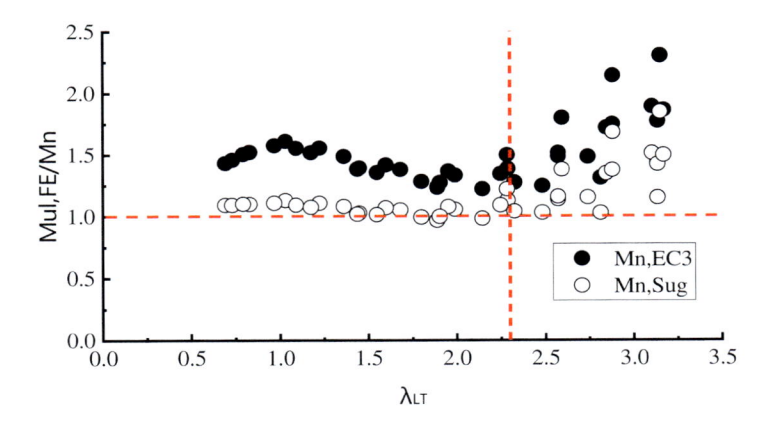

Figure 5.19 Comparison between current ($M_{n,EC3}$) and suggested ($M_{n,Sug}$) design models.

5.5.5 Slenderness limit of inelastic lateral-torsional buckling

It becomes inappropriate to use S690 HSS structurally in girders failing by elastic LTB, as the material cannot be used effectively. Hence, an inelastic slenderness limit ($\lambda_{LT,in}$) deserves to be specified to distinguish these failure modes. On this basis, the relationships between the critical and ultimate BMs versus the slenderness parameters for Groups G1-G5 are presented in Fig. 5.20, from which $\overline{\lambda}_{LT,in}$ can be determined at the intersection of both relationships. Herein, a limit of $\overline{\lambda}_{LT,in} = 2.3$, represented by the red dashed lines, was found approximately to be the end of the inelastic LTB range. As can be noticed, it is quite similar to that of the homogenous CWGs.

5.6   Solved examples

5.6.1 Example #1

Corrugation dimensions of crane carrying, shown in Fig. A1, are considered and the girders height (h_w) is 1000 mm, flange width (b_f) equals 200 mm, and with 6 mm flange thickness (t_f). These calculations were performed on CWG under pure BM with girder length equals 1440 mm (4 waves). This corresponds to a recent work for the authors [53].

The steel used has yield strength equals 235 Mpa. The elastic modulus (E) and Poisson's ratio (υ) are 200,000 MPa and 0.30, respectively.

- Calculating the critical flexure stress according to Moon et al.

$$W_{n1} = \frac{2b_f^2 h_w t_f + b_f h_w^2 t_w}{8b_f t_f + 4h_w t_w} = \frac{2 * 200^2 * 1000 * 6 + 200 * 1000^2 * 6}{8 * 200 * 6 + 4 * 1000 * 6} = 50000$$

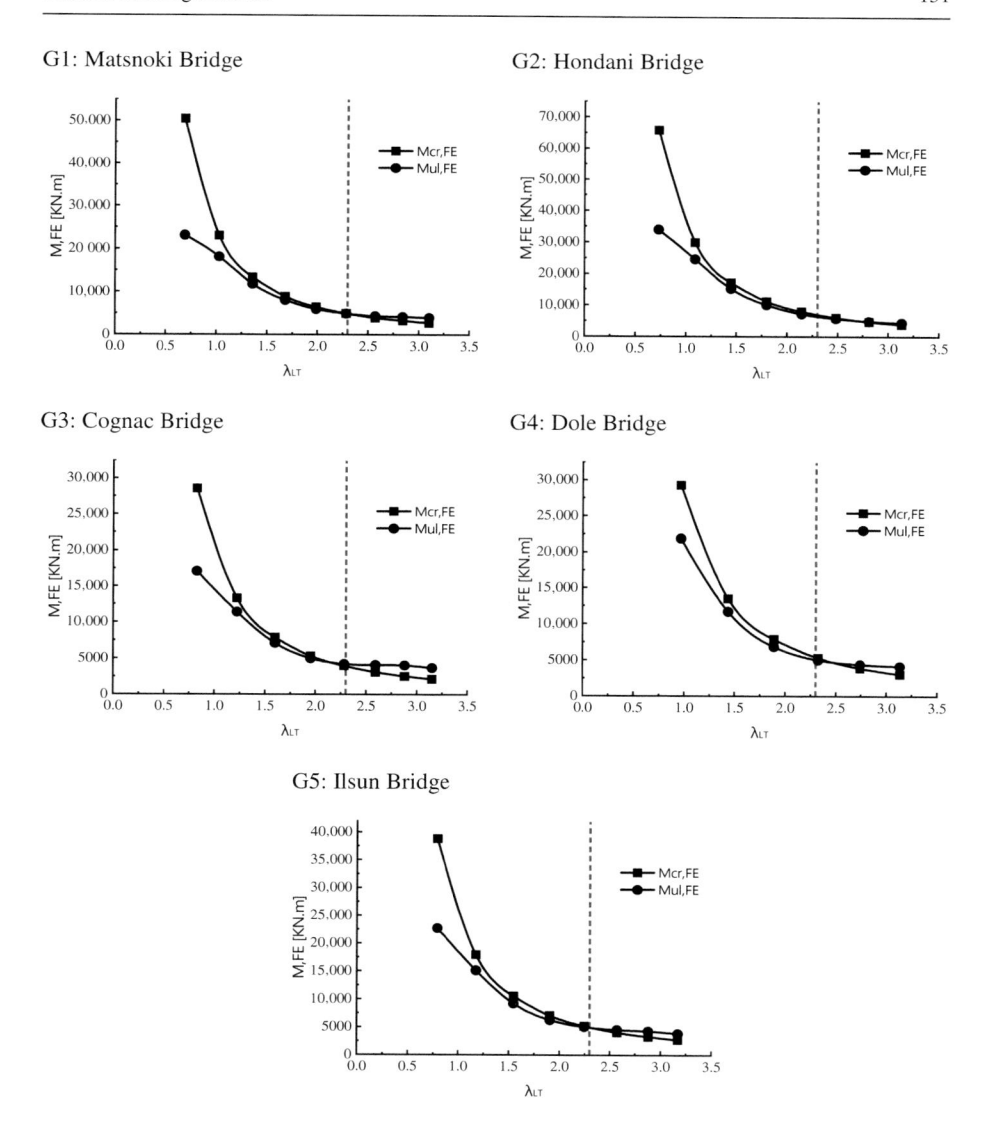

Figure 5.20 Relationships of critical and ultimate bending moments against slenderness parameters for Groups G1-G5—red dashed lines are at $\overline{\lambda}_{LT,in} = 2.3$.

$$W_{n2} = \frac{2b_f^2 h_w t_f + b_f h_w^2 t_w}{8b_f t_f + 4h_w t_w} - \left(\frac{b_f}{4} - \frac{h_r}{2}\right)h_w = \frac{2 * 200^2 * 1000 * 6 + 200 * 1000^2 * 6}{8 * 200 * 6 + 4 * 1000 * 6}$$
$$- \left(\frac{200}{4} - \frac{60}{2}\right) * 1000$$
$$= 11670$$

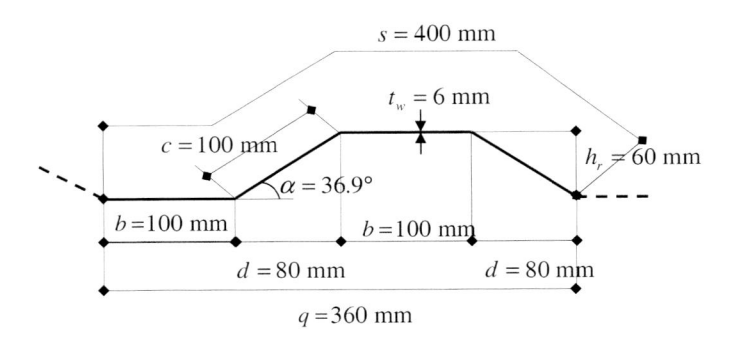

Figure A1 Corrugation configuration.

$$W_{n3} = \frac{2b_f^2 h_w t_f + b_f h_w^2 t_w}{8 b_f t_f + 4 h_w t_w} - \left(\frac{b_f}{4} + \frac{hr}{2}\right) h_w = \frac{2 * 200^2 * 1000 * 6 + 200 * 1000^2 * 6}{8 * 200 * 6 + 4 * 1000 * 6}$$
$$- \left(\frac{200}{4} + \frac{60}{2}\right) * 1000$$
$$= -11670$$

$$W_{n4} = \frac{2b_f^2 h_w t_f + b_f h_w^2 t_w}{8 b_f t_f + 4 h_w t_w} - \frac{1}{2} b_f h_w = \frac{2 * 200^2 * 1000 * 6 + 200 * 1000^2 * 6}{8 * 200 * 6 + 4 * 1000 * 6}$$
$$- \frac{1}{2} * 200 * 1000$$
$$= -50000$$

$$W_{n5} = W_{n4} = -50000$$

$$W_{n6} = W_{n1} = 50000$$

$$d_{\max} = h_r/2 = 60/2 = 30 \, \text{mm}$$

$$d_{aver} = \frac{(2b+d)}{2(b+d)} d_{\max} = \frac{(2 * 100 + 80)}{2 * (100 + 80)} * 30 = 23.34 \, \text{mm}$$

$$L_{1-2} = \frac{b_f}{2} - d_{aver} = \frac{200}{2} - 23.34 = 76.66 \, \text{mm}$$

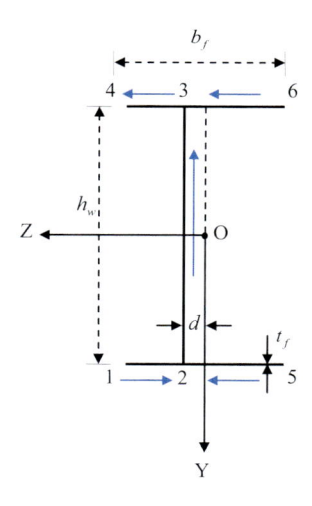

$$L_{2-3} = h_w + t_f = 1000 + 6 = 1006 \text{ mm}$$

$$L_{3-4} = \frac{b_f}{2} - d_{aver} = \frac{200}{2} - 23.34 = 76.66 \text{ mm}$$

$$L_{5-2} = \frac{b_f}{2} + d_{aver} = \frac{200}{2} + 23.34 = 123.34 \text{ mm}$$

$$L_{6-3} = \frac{b_f}{2} + d_{aver} = \frac{200}{2} + 23.34 = 123.34 \text{ mm}$$

$$t_{1-2} = t_f = 6 \text{ mm}$$

$$t_{2-3} = t_w = 6 \text{ mm}$$

$$t_{3-4} = t_f = 6 \text{ mm}$$

$$t_{5-2} = t_f = 6 \text{ mm}$$

$$t_{6-3} = t_f = 6 \text{ mm}$$

$$c_w = \frac{1}{3} \begin{bmatrix} \left[w_{n1}^2 + w_{n1} * w_{n2} + w_{n2}^2\right] * L_{1-2} * t_{1-2} + \left[w_{n2}^2 + w_{n2} * w_{n3} + w_{n3}^2\right] * L_{2-3} * t_{2-3} + \\ \left[w_{n3}^2 + w_{n3} * w_{n4} + w_{n4}^2\right] * L_{3-4} * t_{3-4} + \left[w_{n2}^2 + w_{n2} * w_{n5} + w_{n5}^2\right] * L_{5-2} * t_{5-2} + \\ \left[w_{n3}^2 + w_{n3} * w_{n6} + w_{n6}^2\right] * L_{6-3} * t_{6-3} \end{bmatrix}$$

$$= \frac{1}{3} \begin{bmatrix} \left[50000^2 + 50000 * 11670 + 11670^2\right] * 76.66 * 6 \\ +\left[11670^2 + 11670 * (-11670) + (-11670)^2\right] * 1006 * 6 \\ +\left[(-11670)^2 + (-11670) * (-50000) + (-50000)^2\right] * 76.66 * 6 + \\ \left[11670^2 + 11670 * (-50000) + (-50000)^2\right] * 123.34 * 6 \\ +\left[(-11670)^2 + (-11670) * 50000 + 50000^2\right] * 123.34 * 6 \end{bmatrix}$$

$$= 2.27238E + 12 \, \text{mm}^4$$

$$J_{co} = \frac{1}{3} * (2b_f t_f^3 + h_w t_w^3) = \frac{1}{3} * (2 * 200 * 6^3 + 1000 * 6^3) = 100800 \, \text{mm}^4$$

$$G = \frac{E}{2(1 + \upsilon)} = \frac{200000}{2(1 + 0.3)} = 76923.1 \, \text{Mpa}$$

$$G_{co} = \frac{b + d}{b + c} * G = \frac{100 + 80}{100 + 100} * 76923.1 = 69226.2 \, \text{Mpa}$$

$$I_{y,co} = \frac{t_f b_f^3}{6} = \frac{6 * 200^3}{6} = 8000000 \, \text{mm}^4$$

$$M_{cr,LTB} = \frac{\Pi}{L} \sqrt{EI_{y,co} G_{co} J_{co} \left(1 + \frac{\pi^2}{(L)^2} \frac{EC_w}{G_{co} J_{co}}\right)}$$

$$= \frac{3.14}{1440} \sqrt{200000 * 8000000 * 69226.2 * 100800 * \left(1 + \frac{3.14^2}{1440^2} \frac{200000 * 2.27238E + 12}{69226.2 * 100800}\right)}$$

$$= 4061694946 \, \text{N.mm} = 4062 \, \text{kN.m}$$

- Calculating the ultimate flexure strength according to the modified EC3 equation

$$M_{pl,R} = W_y f_{y,eff} = 200 * 6 * 1000 * 235 = 231656116 \, \text{N.mm} = 232 \, \text{kN.m}$$

$$\lambda_{LT} = \sqrt{\frac{M_{pl,R}}{M_{cr,LTB}}} = \sqrt{\frac{232}{4062}} = 0.24$$

$$\Phi_{LT} = 0.5\left(1 + \alpha_{LT}(\lambda_{LT} - 0.2) + \lambda_{LT}^2\right) = 0.5(1 + 0.13 * (0.24 - 0.2) + 0.24^2) = 0.53$$

$$\chi_{LT} = \frac{1}{\Phi_{LT} + \sqrt{\Phi_{LT}^2 - \overline{\lambda}_{LT}^2}} \leq 1.0 = \frac{1}{0.53 + \sqrt{0.53^2 - 0.24^2}} = 0.995$$

$$M_{EC3, \, \text{modified}} = \chi_{LT} M_{pl,R} = 0.995 * 232 = 230 \, \text{kN.m}$$

$$M_{ul,FE} = 266 \, \text{kN.m}$$

$$\frac{M_{ul,FE}}{M_{EC3,\text{modified}}} = \frac{266}{230} = 1.15 > 1.00 \longrightarrow \text{ok safe}$$

5.6.2 Example #2

Corrugation dimensions of Cognac bridge, shown in Fig. A2, are considered and the girders height (h_w) is 1771 mm, flanges width (b_{f1}, b_{f2}) equal 500 mm and 300 mm, respectively, and with 50 mm flange thicknesses (t_{f1}, t_{f2}). These calculations were

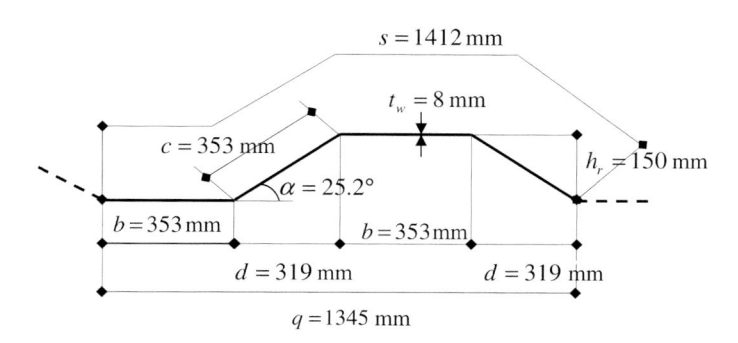

Figure A2 Corrugation configuration of Cognac bridge.

performed on CWG under pure BM with girder length equals 10 waves (13,450 mm). This corresponds to a recent work for the authors [54].

The steel used S460. The elastic modulus (E) and Poisson's ratio (υ) are 200,000MPa and 0.30 respectively.

- Calculating the critical flexure stress according to Ibrahim

$$I_{y1} = \frac{t_{f1}b_{f1}^3}{12} = \frac{50 * 500^3}{12} = 520833333.3 \, \text{mm}^4$$

$$I_{y2} = \frac{t_{f2}b_{f2}^3}{12} = \frac{50 * 300^3}{12} = 112500000 \, \text{mm}^4$$

$$I_{y,c} = I_{y1} + I_{y2} = 520833333.3 + 112500000 = 633333333 \, \text{mm}^4$$

$$\rho = \frac{I_{y1}}{I_{y1} + I_{y2}} = \frac{520833333.3}{520833333.3 + 112500000} = 0.82$$

$$C_{w,mf} = \rho(1-\rho)I_{y,c}h_w^2 = 0.82 * (1-0.82) * 633333333 * 1771^2 = 2.9017E+14 \, \text{mm}^4$$

$$A_{f1} = b_{f1}t_{f1} = 500 * 50 = 25000 \, \text{mm}^2$$

$$A_{f1} = b_{f2}t_{f2} = 300 * 50 = 15000 \, \text{mm}^2$$

$$A_w = (h_w - 0.5t_{f1} - 0.5t_{f2})t_w = (1771 - 0.5 * 50 - 0.5 * 50) * 8 = 13768 \, \text{mm}^2$$

$$A = A_{f1} + A_{f2} + A_w = 25000 + 15000 + 13768 = 53768 \, \text{mm}^2$$

$$\psi = \left(4 - \frac{3h_wt_w}{A}\right) = \left(4 - \frac{3 * 1771 * 8}{53768}\right) = 3.21$$

$$K = 4\rho(1-\rho) + \psi(2\rho - 1)^2 = 4 * 0.82 * (1-0.82) + 3.21 * (2 * 0.82 - 1)^2 = 1.92$$

$$I_w = \frac{h_w^3 t_w}{12} = \frac{1771^3 * 8}{12} = 3.703\text{E} + 09 \text{ mm}^4$$

$$C_{w,mc} = C_{w,mf} + \left(\frac{b + d/3}{b + d}\right) K I_w \left(\frac{h_r}{2}\right)^2 = 12.9017\text{E} + 14 + \left(\frac{353 + 319/3}{353 + 319}\right)$$

$$*1.92 * 3.703\text{E} + 09 * \left(\frac{150}{2}\right)^2 = 3.1756\text{E} + 14 \text{ mm}^4$$

$$J_{co} = \frac{1}{3}(b_{f1}t_{f1}^3 + b_{f2}t_{f2}^3 + h_w t_w^3) = \frac{1}{3}(500*50^3 + 300*50^3 + 1771*8^3) = 33635584 \text{ mm}^4$$

$$G = \frac{E}{2(1 + \upsilon)} = \frac{200000}{2 * (1 + 0.3)} = 76923.1 \text{ Mpa}$$

$$G_{co} = \frac{b + d}{b + c} G = \frac{353 + 319}{353 + 353} * 76923.1 = 73266.2 \text{ Mpa}$$

$$W_1 = \frac{\Pi}{L_b} \sqrt{\frac{EC_{w,mc}}{G_{co}J_{co}}} = \frac{3.14}{13450} * \sqrt{\frac{200000 * 3.1756\text{E} + 14}{73266.2 * 33635584}} = 1.19$$

$$\mu = \frac{A_1}{A_1 + A_2} = \frac{25000}{25000 + 15000} = 0.63$$

$$I_{x,co} = \left(\frac{A_1 A_2}{A_1 + A_2}\right) h_w^2 = \left(\frac{25000 * 15000}{25000 + 15000}\right) * 1771^2 = 29404134375 \text{ mm}^4$$

$$\beta_x = (2\rho - 1)h_w + \frac{I_y}{I_x}(\mu - \rho)h_w = (2 * 0.82 - 1) * 1771 + \frac{633333333}{29404134375} * (0.63 - 0.82) * 1771$$

$$= 1134.30$$

$$W_2 = \frac{\Pi \beta_x}{2L_b} \sqrt{\frac{E I_{yc}}{G_{co}J_{co}}} = \frac{3.14 * 1134.30}{2 * 13450} \sqrt{\frac{200000 * 633333333}{73266.2 * 33635584}} = 0.95$$

$$M_{cr,mc} = \frac{\Pi}{L} \sqrt{E I_{y,co} G_{co} J_{co}} \left[\sqrt{1 + W_1^2 + W_2^2} + w_2\right]$$

$$= \frac{3.14}{13450} * \sqrt{200000 * 633333333 * 73266.2 * 33635584} * \left[\sqrt{1 + 1.19^2 + 0.95^2} + 0.95\right]$$

$$= 1.1417\text{E} + 10 \text{ N.mm} = 11417 \text{ kN.m}$$

- Calculating the ultimate flexure strength according to the modified equation

$$M_{pl,R} = f_y \left(\frac{2A_1 A_2}{A_1 + A_2}\right) h_w = 460 * \left(\frac{2 * 25000 * 15000}{25000 + 15000}\right) * 1771 = 15274875000 \text{ N.mm}$$

$$= 15275 \text{ kN.m}$$

$$\lambda_{LT} = \sqrt{\frac{M_{pl,R}}{M_{cr,mc}}} = \sqrt{\frac{15275}{11417}} = 1.16$$

$$\lambda_{LT} < 1.70$$

$$M_{Rd,sug} = \min \begin{cases} \dfrac{W_{eff} \cdot f_y}{\gamma_{M0}} \\ \dfrac{0.85}{\gamma_{M1}} \cdot \dfrac{\pi}{L}\sqrt{EI_{y,co}G_{co}J_{co}}\left[\sqrt{1 + W_1^2 + W_2^2} + W_2\right] \end{cases}$$

$$= \min \begin{cases} 15275 \text{ kN.m} \\ \dfrac{0.85}{1.0} * 11417 = 9704.45 \text{ kN.m} \end{cases}$$

5.6.3 Example #3

Corrugation dimensions of Maupre bridge, shown in Fig. A3, are considered and the girders height (h_w) is 2650 mm, flange width (b_f) equals 500 mm and with 50 mm flange thickness (t_f). The calculations were performed on CWG under pure BM with girder length equals 14 waves (14700 mm), No.3 in the basic group.

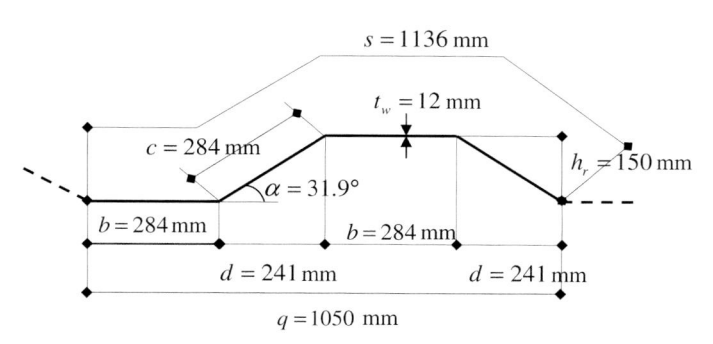

Figure A3 Corrugation configuration of Maupre bridge.

The steel used S460. The elastic modulus (E) and Poisson's ratio (υ) are 200,000MPa and 0.30 respectively.

- Calculating the critical flexure stress according to Moon et al.

$$W_{n1} = \frac{2b_f^2 h_w t_f + b_f h_w^2 t_w}{8b_f t_f + 4h_w t_w} = \frac{2 * 500^2 * 2650 * 50 + 500 * 2650^2 * 12}{8 * 500 * 50 + 4 * 2650 * 12} = 331250$$

$$W_{n2} = \frac{2b_f^2 h_w t_f + b_f h_w^2 t_w}{8b_f t_f + 4h_w t_w} - \left(\frac{b_f}{4} - \frac{h_r}{2}\right)h_w = \frac{2 * 500^2 * 2650 * 50 + 500 * 2650^2 * 12}{8 * 500 * 50 + 4 * 2650 * 12}$$

$$- \left(\frac{500}{4} - \frac{150}{2}\right) * 2650$$

$$= 76562.7$$

$$W_{n3} = \frac{2b_f^2 h_w t_f + b_f h_w^2 t_w}{8b_f t_f + 4h_w t_w} - \left(\frac{b_f}{4} + \frac{hr}{2}\right)h_w = \frac{2 * 500^2 * 2650 * 50 + 500 * 2650^2 * 12}{8 * 500 * 50 + 4 * 2650 * 12}$$

$$-\left(\frac{500}{4} + \frac{150}{2}\right) * 2650$$

$$= -76562.68$$

$$W_{n4} = \frac{2b_f^2 h_w t_f + b_f h_w^2 t_w}{8b_f t_f + 4h_w t_w} - \frac{1}{2}b_f h_w = \frac{2 * 500^2 * 2650 * 50 + 500 * 2650^2 * 12}{8 * 500 * 50 + 4 * 2650 * 12}$$

$$-\frac{1}{2} * 500 * 2650$$

$$= -331250$$

$$W_{n5} = W_{n4} = -331250$$

$$W_{n6} = W_{n1} = 331250$$

$$d_{max} = h_r/2 = 150/2 = 75 \, \text{mm}$$

$$d_{aver} = \frac{(2b + d)}{2(b + d)}d_{max} = \frac{(2 * 284 + 241)}{2 * (284 + 241)} * 75 = 57.78 \, \text{mm}$$

$$L_{1-2} = \frac{b_f}{2} - d_{aver} = \frac{500}{2} - 57.78 = 192.22 \, \text{mm}$$

$$L_{2-3} = h_w + t_f = 2650 + 50 = 2700 \, \text{mm}$$

$$L_{3-4} = \frac{b_f}{2} - d_{aver} = \frac{500}{2} - 57.78 = 192.22 \, \text{mm}$$

$$L_{5-2} = \frac{b_f}{2} + d_{aver} = \frac{500}{2} + 57.78 = 307.78 \, \text{mm}$$

$$L_{6-3} = \frac{b_f}{2} + d_{aver} = \frac{500}{2} + 57.78 = 307.78 \, \text{mm}$$

$$t_{1-2} = t_f = 50 \, \text{mm}$$

$$t_{2-3} = t_w = 12 \, \text{mm}$$

$$t_{3-4} = t_f = 50 \, \text{mm}$$

$$t_{5-2} = t_f = 50 \, \text{mm}$$

$$t_{6-3} = t_f = 50 \, \text{mm}$$

$$c_w = \frac{1}{3} \begin{bmatrix} \left[w_{n1}^2 + w_{n1} * w_{n2} + w_{n2}^2\right] * L_{1-2} * t_{1-2} + \left[w_{n2}^2 + w_{n2} * w_{n3} + w_{n3}^2\right] * L_{2-3} * t_{2-3} \\ + \left[w_{n3}^2 + w_{n3} * w_{n4} + w_{n4}^2\right] * L_{3-4} * t_{3-4} + \left[w_{n2}^2 + w_{n2} * w_{n5} + w_{n5}^2\right] * L_{5-2} * t_{5-2} \\ + \left[w_{n3}^2 + w_{n3} * w_{n6} + w_{n6}^2\right] * L_{6-3} * t_{6-3} \end{bmatrix}$$

$$= \frac{1}{3} \begin{bmatrix} \left[331250^2 + 331250 * 76562.7 + 76562.7^2\right] * 192.22 * 50 \\ + \begin{bmatrix} 76562.7^2 + 76562.7 * (-76562.68) \\ + (-76562.68)^2 \end{bmatrix} * 2700 * 12 \\ + \left[(-76562.68)^2 + (-76562.68) * (-331250) + (-331250)^2\right] * 192.22 * 50 \\ + \left[76562.7^2 + 76562.7 * (-331250) + (-331250)^2\right] * 307.78 * 50 \\ + \left[(-76562.68)^2 + (-76562.68) * 331250 + 331250^2\right] * 307.78 * 50 \end{bmatrix}$$

$$= 1.89091E + 15 \, \text{mm}^4$$

$$J_{co} = \frac{1}{3} * (2b_f t_f^3 + h_w t_w^3) = \frac{1}{3} * (2 * 500 * 50^3 + 2650 * 12^3) = 43193067 \, \text{mm}^4$$

$$G = \frac{E}{2(1 + \upsilon)} = \frac{200000}{2(1 + 0.3)} = 76923.1 \, \text{Mpa}$$

$$G_{co} = \frac{b + d}{b + c} * G = \frac{284 + 241}{284 + 284} * 76923.1 = 71120 \, \text{Mpa}$$

$$I_{y,co} = \frac{t_f b_f^3}{6} = \frac{50 * 500^3}{6} = 1041666667 \, \text{mm}^4$$

$$M_{cr,LTB} = \frac{\Pi}{L} \sqrt{EI_{y,co} G_{co} J_{co} \left(1 + \frac{\pi^2}{(L)^2} \frac{EC_w}{G_{co} J_{co}}\right)}$$

$$= \frac{3.14}{14700} \sqrt{200000 * 1041666667 * 71120 * 43193067 * \left(1 + \frac{3.14^2}{14700^2} \frac{200000 * 1.891E + 15}{71120 * 43193067}\right)}$$

$$= 1.3893E + 10 \, \text{N.mm} = 13893 \, \text{kN.m}$$

- Calculating the ultimate flexure strength according to the modified equation

$$M_{p1,R} = W_y f_y = 500 * 50 * 2650 * 460 = 30475E + 10^6 \, \text{N.mm} = 30475 \, \text{kN.m}$$

$$\lambda_{LT} = \sqrt{\frac{M_{pl,R}}{M_{cr,LTB}}} = \sqrt{\frac{30475}{13893}} = 1.48$$

$$\Phi_{LT} = 0.5\left(1 + \alpha_{LT}(\lambda_{LT} - 0.2) + \lambda_{LT}^2\right) = 0.5(1 + 0.13 * (1.48 - 0.2) + 1.48^2) = 1.68$$

$$\chi_{LT} = \frac{1}{\Phi_{LT} + \sqrt{\Phi_{LT}^2 - \bar{\lambda}_{LT}^2}} \leq 1.0 = \frac{1}{1.68 + \sqrt{1.68^2 - 1.48^2}} = 0.404$$

$$M_{EC3, \text{modified}} = \chi_{LT} M_{p1,R} = 0.404 * 30475 = 12323 \, \text{kN.m}$$

$$M_{ul,FE} = 12843 \, \text{kN.m}$$

$$\frac{M_{ul,FE}}{M_{EC3,\text{modified}}} = \frac{12843}{12323} = 1.04 > 1.00 \longrightarrow \text{ok safe}$$

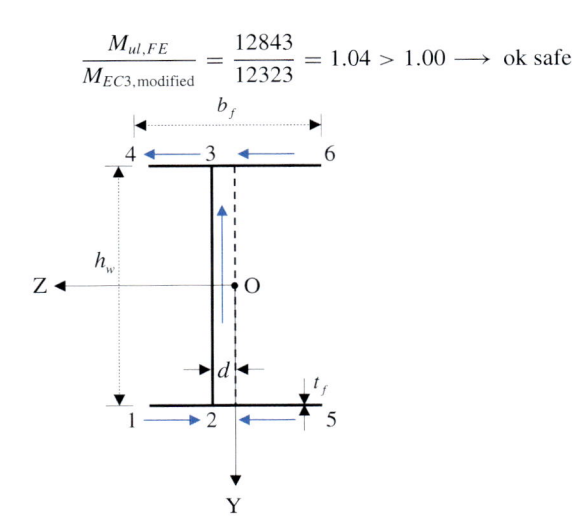

References

[1] Q.A. Hasan, W.H.W. Badaruzzaman, A.W. Al-Zand, A.A. Mutalib, "The state of the art of steel and steel concrete composite straight plate girder bridges", Thin-Walled Struct. 119 (2017) 988–1020.

[2] S. Ikeda, M. Sakurada, "Development of hybrid prestressed concrete bridges with corrugated steel web plates", in: 30th Conference on Our World in Concrete & Structures, Singapore, 2005.

[3] J. Moon, J. Yi, B.H. Choi, H. Lee, "Shear strength and design of trapezoidally corrugated steel webs", J. Constr. Steel Res. 65 (2009) 1198–1205.

[4] M.F. Hassanein, A.A. Elkawas, A.M. El Hadidy, M. Elchalakani, "Shear analysis and design of high-strength steel corrugated web girders for bridge design", Eng. Struct. 146 (2017) 18–33.

[5] J. Lindner, "Lateral torsional buckling of beams with trapezoidally corrugated webs", Stab. Steel Struct. Budapest, Hungary (1990) 305–308.

[6] R. Sause, H.H. Abbas, W.G. Wassef, R.G. Driver, M. Elgaaly, "Corrugated web girder shape and strength criteria", Civil and Environmental Engineering, Lehigh University, ATLSS Rep. (03-18) (2003).

[7] E.Y. Sayed-Ahmed, "Lateral torsion-flexure buckling of corrugated web steel girders", Proc. Inst. Civil Eng., Struct. Build. 158 (1) (2005) 53–69.

[8] J. Moon, J-W. Yi, B.H. Choi, H-E. Lee, "Lateral-torsional buckling of I-girder with corrugated webs under uniform bending", Thin-Walled Struct. 47 (2009) 21–30.

[9] N.D. Nguyen, S.N. Kim, S-R. Han, Y-J. Kang, "Elastic lateral-torsional buckling strength of I-girder with trapezoidal web corrugations using a new warping constant under uniform moment", Eng. Struct. 32 (2010) 2157–2165.

[10] J. Moon, N-H. Lim, H-E. Lee, "Moment gradient correction factor and inelastic flexural-torsional buckling of I-girder with corrugated steel webs", Thin-Walled Struct. 62 (2013) 18–27.

[11] E.Y. Sayed-Ahmed, "Design aspects of steel I-girders with corrugated webs", Electron. J. Struct. Eng. 7 (2007) 27–40.

[12] S.P. Timoshenko, J.M. Gere, "Theory of Elastic Stability", 2nd edition, McGraw-Hill, London, 1961.

[13] L. Euler, "Sur La Force Des Colonnes", Memoires Academic Royale des Sciences et Belle Lettres, 13, 1759, pp. 309–318.

[14] B. Saint-Venant, Memoire Sur La Torsion Des Prismes", "Memoires des Savants Etrangers", XIV, 1855, pp. 233–560.

[15] A.G.M. Michell, "Elastic stability of long beams under transverse forces", Philos. Mag. 1 (48) (1899) 298–309.

[16] L. Prandtl, PhD thesis, Munich, 1899.

[17] A.S.J. Foster, "Stability and design of steel beams in the strain-hardening range", PhD thesis, Department of Civil and Environmental Engineering, Imperial College London, 2014.

[18] V.Z. Vlasov, Тонкостенные упругие стерЖни, Gosudatvenoe izdavateljstvo fiziko-matematiceskoj literaturi, Moscau (1940).

[19] S.P. Timoshenko, The collected papers of Stephen P. Timoshenko, chapter Einige Stabilitaetsprobleme der Elastizitaetstheorie, McGraw-Hill, New York, 1953, pp. 1–50.

[20] S.P. Timoshenko, "The Collected Papers of Stephen P. Timoshenko", chapter Sur la stabilite des systemes elastiques, McGraw-Hill, 1953, pp. 92–224.

[21] M. Larsson, J. Persson, "Lateral-torsional buckling of steel girders with trapezoidally corrugated webs", MSc thesis, Chalmers University of Technology, Sweden, 2013.

[22] S.A. Ibrahim, "Lateral torsional buckling strength of unsymmetrical plate girders with corrugated webs", Eng. Struct. 81 (2014) 123–134.

[23] H.H. Abbas, "Analysis and design of corrugated web I-girders for bridges using high performance steel", Department of Civil and Environmental Engineering, Lehigh University, Bethlehem, PA, 2003.

[24] J.Y.R. Liew, M. Xiong, D. Xiong, "Design of concrete filled tubular beam-columns with high strength steel and concrete", Structures 8 (2) (2016) 213–226.

[25] G.-Q. Li, H. Lyu, C. Zhang, "Post-fire mechanical properties of high strength Q690 structural steel", J. Constr. Steel Res. 132 (2017) 108–116.

[26] P.V. Nidheesh, M.S. Kumar, "An overview of environmental sustainability in cement and steel production", J. Cleaner Prod. 231 (2019) 856–871.

[27] WSA: World Steel Association"The Three Rs of Sustainable Steel", World Steel Association, Brussels, Belgium, 2010.

[28] F. Schröter, "Trends of using high-strength steel for heavy steel structures", in: MA Giejowski, A. Kozowski, L. lczka, J. Zióko (Eds.), Progress in Steel, Composite and Aluminium Structures, CRC Press, Boca Raton, Florida, 2006, pp. 292–293.

[29] R. Sause, "Corrugated web girder fabrication", Pennsylvania Innovative High Performance Steel Bridge Demonstration Project, ATLSS Report No. 03-19, 2003.

[30] M.A. Bradford, X. Liu, "Flexural-torsional buckling of high-strength steel beams", J. Constr. Steel Res. 124 (2016) 122–131.

[31] B. Somodi, B. Kövesdi, "Flexural buckling resistance of welded HSS box section members", Thin-Walled Struct. 119 (2017) 266–281.

[32] EN 1993-1-1Eurocode 3: Design of steel structures - Part 1-1: General rules and rules for buildings, CEN, 2004.

[33] B. Somodi, B. Kövesdi, "Residual stress measurements on cold-formed HSS hollow section columns", J. Constr. Steel Res. 128 (2017) 706–720.

[34] A.A. Elkawas, M.F. Hassanein, M. Elchalakani, "Lateral-torsional buckling strength and behaviour of high-strength steel corrugated web girders for bridge construction", Thin-Walled Struct. 122 (2018) 112–123.

[35] R.G. Driver, H.H. Abbas, R. Sause, "Shear behavior of corrugated web bridge girders", J. Struct. Eng., ASCE 132 (2) (2006) 195–203.

[36] A.A. Elkawas, M.F. Hassanein, M.H. El-Boghdadi, "Numerical investigation on the nonlinear shear behaviour of high-strength steel tapered corrugated web bridge girders", Eng. Struct. 134 (2017) 358–375.

[37] E. Gogou, "Use of high strength steel grades for economical bridge design", Master thesis study, Delft University of Technology Iv-Infra, Amsterdam, 2012.

[38] H.-P. Günther, J. Raoul, IABSE: International Association for Bridge and Structural Engineering, "Use and application of high performance steels for steel structures", Struct. Eng. Doc. 8 (2005).

[39] R.W. Frost, C.G. Schilling, "Behavior of hybrid beams subjected to static loads", J. Struct. Div., ASCE 90 (3) (1964) 55–88.

[40] M. Shokouhian, Y. Shi, M. Head, "Interactive buckling failure modes of hybrid steel flexural members", Eng. Struct. 125 (2016) 153–166.

[41] C.-S. Wang, L. Duan, Y.F. Chen, S.-C. Wang, "Flexural behavior and ductility of hybrid high performance steel I-Girders", J. Constr. Steel Res. 125 (2016) 1–14.

[42] Y.-B. Shao, Y.-M. Zhang, M.F. Hassanein, "Strength and behaviour of laterally-unrestrained S690 high-strength steel hybrid girders with corrugated webs", Thin-Walled Struct. 150 (2020) 106688.

[43] ABAQUSStandard User's Manual, 2013 The Abaqus Software is a product of Dassault Systèmes Simulia Corp., Providence, RI, USA Dassault Systèmes, Version 6.13.

[44] E. Zevallos, M.F. Hassanein, E. Real, E. Mirambell, "Shear evaluation of tapered bridge girder panels with steel corrugated webs near the supports of continuous bridges", Eng. Struct. 113 (2016) 149–159.

[45] J. Yi, H. Gil, K. Youm, H. Lee, "Interactive shear buckling behavior of trapezoidally corrugated steel webs", Eng. Struct. 30 (2008) 1659–1666.

[46] AISC 360-16"Load and Resistance Factor Design Specification, for Structural Steel Buildings", American Institute of Steel Construction, Chicago, IL, 2016.

[47] Y. Sun, A. He, Y. Liang, O. Zhao, "In-plane bending behaviour and capacities of S690 high strength steel welded I-Section beams", J. Constr. Steel Res. 162 (2019) 105741.

[48] R.W. Hamilton, "Behavior of welded girder with corrugated webs", Ph.D. thesis, University of Maine, Orono, Maine, 1993.

[49] J. He, Y. Liu, A. Chen, T. Yoda, "Mechanical behavior and analysis of composite bridges with corrugated steel webs: state-of-the-art", Int. J. Steel Struct. 12 (3) (2012) 321–338.

[50] EN 1993-1-12, "Eurocode 3: Design of steel structures - Part 1-12: Additional rules for the extension of EN 1993 up to steel grades S 700", CEN (2009).

[51] B. Jáger, L. Dunai, B. Kövesdi, "Flange buckling behavior of girders with corrugated web part II: numerical study and design method development", Thin-Walled Struct. 118 (2017) 181–195.

[52] M. Gkantou, M. Theofanous, C. Baniotopoulos, "Plastic design of hot-finished high strength steel continuous beams", Thin-Walled Struct. 133 (2018) 85–95.

[53] M.F. Hassanein, A.A. Elkawas, Y.-B. Shao, "Assessment of the suitability of eurocode design model for corrugated web girders with slender flanges", Structures 27 (2020) 1551–1569.

[54] M.F. Hassanein, A.A. Elkawas, Y.-B. Shao, M. Elchalakani, A.M. El Hadidy, "Lateral-torsional buckling behaviour of mono-symmetric S460 corrugated web bridge girder", Thin-Walled Structures 153 (2020), 106803.

Stress analysis of I-girders with concrete-filled tubular flange and corrugated web

6.1 General

Steel I-girder has been widely used in building and bridge engineering due to its ease construction and high flexural strength. However, the design of this girder is limited in the dimension of both flange and web due to the buckling problem. Conventional I-girder includes three parts: top flange, bottom flange, and a web. Generally, both flanges and web are flat steel plate in an I-girder, and rolled H-shape steel is mostly used in the I-girder, as shown in Fig. 6.1. Conventional I-girder is sensitive to buckling because of two reasons: (1) firstly, the H-shape cross-section is an open section showing weakness in resisting torsion action. Hence, an I-girder with large height is sensitive to global buckling because this failure mode is dominated by lateral flexural-torsional deformation. (2) Secondly, large width/thickness ratio of the flange and the height/thickness ratio of the web, which is necessary for I-girder to gain a high flexural strength, generally lead to local buckling in them.

Due to the buckling problem as mentioned above, the design size of a flat-plate I-girder is usually limited. To improve the global buckling performance of the I-girder, it is efficient to increase the out-of-plane stiffness of the flanges, including torsional stiffness and lateral flexural stiffness. An innovative method is to replace the flat-plate flange with closed section tubes, such as rectangular, circular, triangular, or pentagonal tube, as illustrated in Fig. 6.2. Researchers (Pi and Trahair [1]; Kim and Sause [2]) focused on this method and studied the global buckling of H-shape beams with tubular flange, and they presented theoretical analysis for predicting the buckling strength. Anapayan et al. [3,4] experimentally studied the flexural strength and global buckling of a channel beam with tubular flanges, and they presented improved design methods for this type of beam against lateral distortional buckling.

Although tubular flange with hollow section is advantageous in resisting lateral deformation including lateral-torsional and flexural deformation, it is sensitive to local buckling because of its thin-walled characteristics, especially when the tubular flange is subjected to lateral local compression. To avoid such disadvantage, Sause et al. [5] presented to fill concrete into the tubular flanges to form composite flanges, and typical concrete-filled tubular flanges (CFTFs) are shown in Fig. 6.3. The improved torsional stiffness of the I-girder with CFTFs can prevent the global buckling efficiently. The flexural strength of a curved I-girder was improved greatly with the help of CFTFs [6]. Gao et al. [7] suggested to replace rectangular or circular flange with pentagonal flange. A pentagonal flange has large height and large volume compared to a rectangular or circular one. It has two advantages: firstly, a larger height of the tubular flange can

Behavior and Design of Trapezoidally Corrugated Web Girders for Bridge Construction: Recent Advances.
DOI: https://doi.org/10.1016/B978-0-323-88437-2.00003-4

Figure 6.1 Rolled H-shape steel.

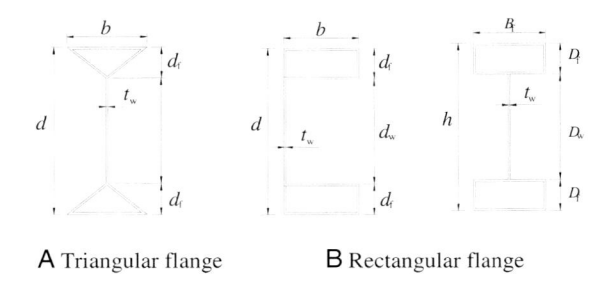

A Triangular flange   **B** Rectangular flange

Figure 6.2 Tubular flange beams.

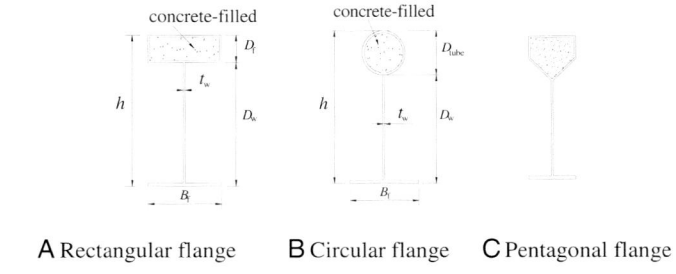

A Rectangular flange   **B** Circular flange   **C** Pentagonal flange

Figure 6.3 Girders with concrete-filled tubular flanges.

reduce the height of the web, and hence the height-to-thickness ratio of the web is decreased to improve the resistance of the web to shear buckling. Secondly, a larger volume of the tubular flange can hold more infilled concrete, which is useful to improve the lateral stiffness of the girder and to increase the strength of global buckling.

Because CFTFs can improve the resistance to global buckling, the I-girders with CFTFs are expected to be designed in large height to improve the flexural yield strength. However, a large height produces local buckling easily to flat-plate web. To avoid local buckling of the web, conventional method is to place stiffeners, including

 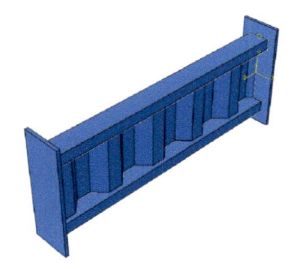

A Top CFTF and bottom flat plate flange **B** Top and bottom CFTFs

Figure 6.4 I-girder with CFTF and CW. *CFTF*, concrete-filled tubular flange; *CW*, corrugated web.

longitudinal and transverse stiffeners, to the web through welding. However, welding stiffeners to the web of an I-girder has two disadvantages: (1) residual stress and possibly large residual deformation may be produced, and (2) self-weight of the girder increases. Corrugated web (CW), as studied in previous sessions, is very efficient to solve the above two problems because it has much higher out-of-plane flexural stiffness compared to flat-plate web. Combining the advantages of CFTF and CW, a new type of I-girders with CFTF and CW was presented by Shao and Wang [8,9], Wang and Shao [10], and Wang et al. [11] (Fig. 6.4). Because the CFTF is mainly used to improve the resistance to global buckling, only the flange under compression in an I-girder is necessary to be CFTF, i.e., only the top flange is CFTF when the I-girder is hinged at both ends while both top and bottom flanges are CFTFs when the I-girder is fixed at both ends.

Compared to flat plate flanges, CFTFs definitely have much larger flexural and torsional stiffnesses, and they can provide much stronger constraints to the web. Fixed boundary conditions can be assumed at the connection of the CFTF and the flat-plate web or the CW. Prediction on global buckling analysis of the I-girders with CFTF and CW is similar to previous studies on global buckling analysis of tubular flange I-girders [1–4] by replacing the flexural and torsional stiffnesses of hollow tubular flanges with corresponding values of CFTFs. Similarly, prediction on local buckling of the I-girders with CFTF and CW can also use the same presented method by Hassanein and Kharoob [12,13] and Wang et al. [11]. Besides buckling problem methods for analyzing stress distribution on the cross-section of the presented new I-girders and for predicting the flexural yielding strength are necessary to be provided for design purposes.

6.2 Normal stress in flange

To provide an analytical method to calculate the stress distribution in an I-girders with CFTF and CW, the following assumptions are presented:

(**1**) There is no relative slip between the steel tube and the infilled concrete in the top flange under compression. This assumption can be satisfied when end plates are welded to both ends of the CFTFs to prevent such slip.

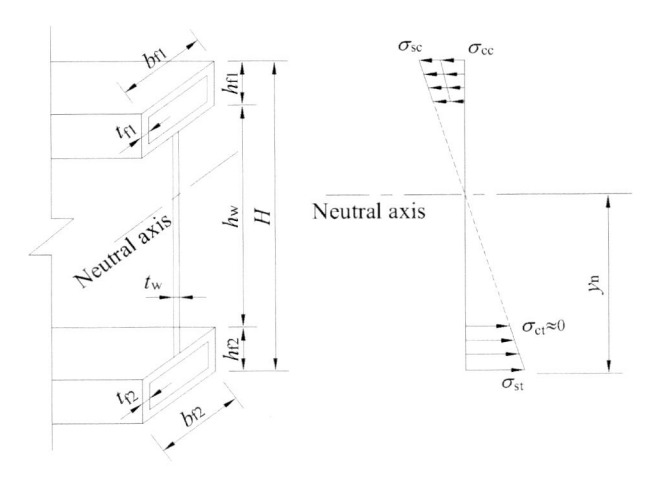

Figure 6.5 Normal stress distribution in flanges.

(2) Plane assumption is still satisfied on the cross-section of the I-girders. This assumption can be satisfied when the flexural stiffness of the endplates is much larger than that of the CFTFs. When there are no endplates connected to the CFTFs and the CW, whether this assumption is satisfied is necessary to be analyzed through experimental test or numerical simulation. Based on this assumption, the strain is linearly distributed along with the height on the cross-section in linear and elastic state.

(3) The tensile strength of the concrete can be neglected. Hence, the contribution of the infilled concrete in the tensile flange (i.e., the bottom CFTF) to the flexural strength is not considered.

(4) The bending moment is only sustained by the two flanges, which indicates that the CW has no contribution to tension or compression. This assumption has been verified from experimental tests. However, the shear force on the cross-section is sustained by both the CFTFs and the CW.

According to assumption (4), normal stress is considered to only exist in flanges. Based on assumptions (1–3), the strains on the cross-section have a linear distribution in the flanges, as illustrated in Fig. 6.5. According to assumption (3), there is no normal tensile stress in the infilled concrete in the bottom tubular flange ($\sigma_{ct} \approx 0$), and such normal tensile stress only exists in the steel tube (σ_{st}). The top CFTF is subjected to compression, and normal compressive stress exists in both infilled concrete (σ_{cc}) and steel tube (σ_{sc}).

For an I-girder with CFTFs and CW subjected to a bending moment M, the following equation can be easily obtained:

$$M = \int_A y \cdot \sigma \cdot dA = \int_{A_{st}} y \cdot \sigma_{st} \cdot dA + \int_{A_{sc}} y \cdot \sigma_{sc} \cdot dA + \int_{A_{cc}} y \cdot \sigma_{cc} \cdot dA \qquad (6.1)$$

The stress components are calculated from the corresponding strains in linear and elastic stages as follows:

$$\begin{cases} \sigma_{st} = E_s \cdot \varepsilon_{st} \\ \sigma_{sc} = E_s \cdot \varepsilon_{sc} \\ \sigma_{cc} = E_c \cdot \varepsilon_{cc} \end{cases}$$

$$(6.2)$$

where ε_{st}, ε_{sc}, and ε_{cc} are the normal strains corresponding to the stresses σ_{st}, σ_{sc}, and σ_{cc}, respectively. E_s and E_c are the elastic moduli for steel and concrete materials, respectively.

The strain at any point on the cross-section of the flanges (ε) is determined by the curvature ($1/\rho$) and the distance to the neutral axial in the height direction on the cross-section (y) as follows:

$$\varepsilon = \frac{y}{\rho}$$

$$(6.3)$$

Substituting Eqs. (6.2 and 6.3) into Eq. (6.1), the following equation can be obtained:

$$M_1 = \int_{A_{st}} E_s \cdot \frac{y^2}{\rho} \cdot dA + \int_{A_{sc}} E_s \cdot \frac{y^2}{\rho} \cdot dA + \int_{A_{cc}} E_c \cdot \frac{y^2}{\rho} \cdot dA = \frac{E_s}{\rho} \cdot (I_{st} + I_{sc}) + \frac{E_c}{\rho} \cdot I_{cc}$$

$$(6.4)$$

where I_{st}, I_{sc}, and I_{cc} are moment of inertia for bottom steel tube flange, for top steel tube flange, and for infilled compressive concrete, respectively, and their expressions are listed as follows, respectively:

$$I_{st} = \frac{1}{12}\left[b_{f2} \cdot h_{f2}^3 - (b_{f2} - 2 \cdot t_{f2}) \cdot (h_{f2} - 2 \cdot t_{f2})^3\right]$$
$$+ \left[b_{f2} \cdot h_{f2} - (b_{f2} - 2 \cdot t_{f2}) \cdot (h_{f2} - 2 \cdot t_{f2})\right] \cdot \left(y_n - \frac{h_{f2}}{2}\right)^2$$
$$I_{sc} = \frac{1}{12}\left[b_{f1} \cdot h_{f1}^3 - (b_{f1} - 2 \cdot t_{f1}) \cdot (h_{f1} - 2 \cdot t_{f1})^3\right]$$
$$+ \left[b_{f1} \cdot h_{f1} - (b_{f1} - 2 \cdot t_{f1}) \cdot (h_{f1} - 2 \cdot t_{f1})\right] \cdot \left(H - y_n - \frac{h_{f1}}{2}\right)^2$$
$$I_{cc} = \frac{1}{12} \cdot (b_{f1} - 2 \cdot t_{f1}) \cdot (h_{f1} - 2 \cdot t_{f1})^3 + (b_{f1} - 2 \cdot t_{f1}) \cdot \left(h_{f1} - 2 \cdot t_{f1}\right) \cdot$$
$$\left(H - y_n - \frac{h_{f1}}{2}\right)^2$$

where b_{f1} and b_{f2} are the widths of the top and bottom CFTFs, respectively; t_{f1} and t_{f2} are the thicknesses of steel tubes on the top and bottom flanges, respectively; h_{f1} and h_{f2} are the heights of steel tubes on the top and bottom flanges, respectively; H is the height of the new girder. All the quantities listed in the above equations and expressions can be found in Fig. 6.5.

The normal stress in the steel tubes can be calculated from the following equation by combining Eqs. (6.2–6.4) together:

$$\sigma_s = E_s \cdot \varepsilon = E_s \cdot \frac{y}{\rho} = E_s \cdot \frac{y \cdot M_1}{E_s \cdot (I_{st} + I_{sc}) + E_c \cdot I_{cc}} = \frac{M_1 \cdot y}{(I_{st} + I_{sc}) + \frac{E_c}{E_s} \cdot I_{cc}} = \frac{M_1 \cdot y}{I_e}$$

(6.5)

where I_e is an equivalent moment of inertia of the cross-section ($I_e = (I_{st} + I_{sc}) + \frac{E_c}{E_s} \cdot I_{cc}$).

Similarly, the stress in the concrete can be also obtained as follows:

$$\sigma_c = E_c \cdot \varepsilon = E_c \cdot \frac{y}{\rho} = E_c \cdot \frac{y \cdot M_1}{E_s \cdot (I_{st} + I_{sc}) + E_c \cdot I_{cc}} = \frac{M_1 \cdot y}{\frac{E_s}{E_c} \cdot (I_{st} + I_{sc}) + I_{cc}} = \frac{M_1 \cdot y}{\frac{E_s}{E_c} \cdot I_e}$$

(6.6)

Eqs. (6.5) and (6.6) can be used to calculate the normal stresses in the steel tubes and in the compressive concrete in top flange. However, such equations can be used only after the location of the neutral axis (y_n) is determined. The neutral axis is determined based on the fact that the axial force on the cross-section is zero, i.e., the following equation must be satisfied:

$$N = \int_A \sigma \, dA = N_{sc} + N_{cc} + N_{st} = 0$$

(6.7)

where N_{sc}, N_{cc}, N_{st} are the axial forces of the steel tube on the top flange, of the infilled concrete on the top flange, and of the steel tube on the bottom flange respectively, and they are calculated from the following equations:

$$N_{sc} = \int_{A_{sc}} \sigma \cdot dA = \int_{A_{sc}} \frac{M_1 \cdot y}{I_e} \cdot dA = \frac{M_1}{I_e} \int_{A_{sc}} y \cdot dA = \frac{M_1}{I_e} \cdot A_{sc} \cdot \left(H - y_n - \frac{h_{f1}}{2} \right)$$

(6.8)

$$N_{cc} = \int_{A_{cc}} \sigma \cdot dA = \frac{M_1}{\frac{E_s}{E_c} I_e} \int_{A_{cc}} y \cdot dA = \frac{M_1}{\frac{E_s}{E_c} I_e} \cdot A_{cc} \cdot \left(H - y_n - \frac{h_{f1}}{2} \right)$$

(6.9)

$$N_{st} = \int_{A_{st}} \sigma \cdot dA = \frac{M_1}{I_e} \int_{A_{st}} y \cdot dA = \frac{M_1}{I_e} \cdot A_{st} \cdot \left(y_n - \frac{h_{f2}}{2} \right)$$

(6.10)

where, A_{sc}, A_{cc}, A_{st} are the corresponding cross-section areas of the steel tube on the top flange ($A_{sc} = b_{f1} \cdot h_{f1} - (b_{f1}\text{-}2 \cdot t_{f1}) \cdot (h_{f1} - 2 \cdot t_{f1})$), of the infilled concrete on the top flange ($A_{cc} = (b_{f1}\text{-}2 \cdot t_{f1}) \cdot (h_{f1} - 2 \cdot t_{f1})$), and of the steel tube on the bottom flange ($A_{st} = b_{f2} \cdot h_{f2} - (b_{f2}\text{-}2 \cdot t_{f2}) \cdot (h_{f2} - 2 \cdot t_{f2})$) respectively.

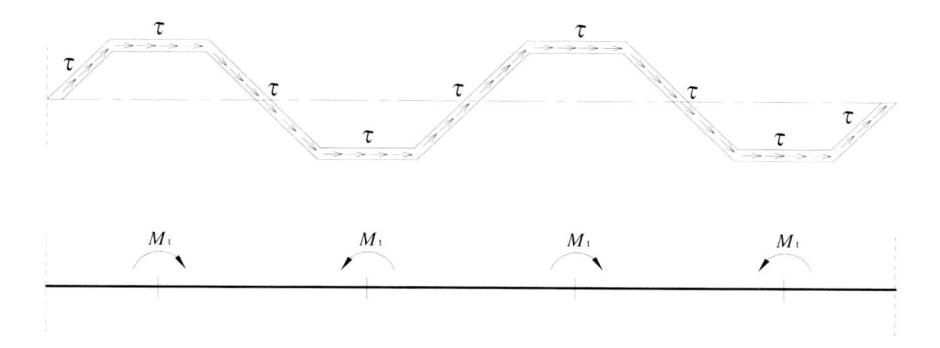

Figure 6.6 Shear stress flow and transverse bending moment.

Substituting Eqs. (6.8)–(6.10) into Eq. (6.7), the neutral axial location can be determined from the following equation:

$$y_n = \frac{\frac{h_{f2}}{2} \cdot \eta_1 + \left(H - \frac{h_{f1}}{2}\right)}{\eta_1 + 1}$$ (6.11)

where

$$\eta_1 = \frac{A_{st}}{A_{sc} + \frac{E_c}{E_s} \cdot A_{cc}}$$

After the neutral axis is determined from Eq. (6.11), the normal stress produced by the bending moment M can be estimated from Eqs. (6.5) and (6.6). However, the flanges are also subjected to transverse bending moment (M_t) produced by the shear stress flow in the CW as illustrated in Fig. 6.6. Abbas et al. [14,15] provided the calculation of M_t in the following equation by using a fictitious load method:

$$M_t = \frac{2V_y}{H}(-A + \xi A_L)$$ (6.12)

where V_y is the shear force at the given cross-section, A is the accumulated area under the corrugation profile, as shown in Fig. 6.7; A_L is the accumulated area under the corrugation profile along the entire span of the I-girder; ξ is a ratio of the given length to the span, i.e., $\xi = z/L$ (z is the distance to the girder's end in the length direction of the girder, and L is the girder's length).

Once the transverse bending moment M_t is determined from Eq. (6.12), the transverse normal stresses in the top CFTF are then calculated from the following equations:

$$\sigma'_{sc} = \frac{M_t x}{I_{ts1} + E_c/E_s \cdot I_{tc}} = \frac{M_t x}{I_{te}}$$ (6.13)

$$\sigma'_{cc} = \frac{M_t x}{E_s/E_c \cdot I_{ts1} + I_{tc}} = \frac{M_t x}{E_s/E_c \cdot I_{te}}$$ (6.14)

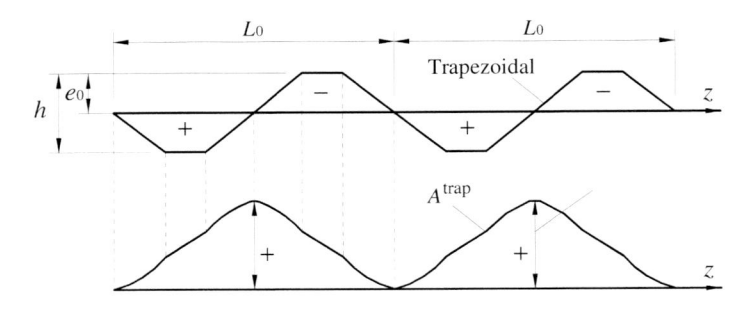

Figure 6.7 Accumulated area A.

where σ_{sc}' and σ_{cc}' are the transverse normal stresses in the steel tube and in the concrete, respectively; I_{ts1} and I_{tc} are the transverse moments of inertia for the steel tube and for the concrete, respectively; I_{te} is the equivalent transverse moment of inertia for the CFTF, $I_{te} = I_{ts1} + E_c/E_s \cdot I_{tc}$.

The transverse normal stress in the bottom flange is calculated from the following equation:

$$\sigma_{st}' = \frac{M_t x}{I_{ts2}} \tag{6.15}$$

where I_{ts2} are the transverse moments of inertia for the bottom steel tube.

Based on Eqs. (6.5, 6.6, 6.13–6.15), the normal stresses in the steel tube and in the concrete can be calculated from the following equation:

$$\begin{cases} \sigma_{sc} = \sigma_s + \sigma_{sc}' = \dfrac{M_x y}{I_e} + \dfrac{M_t x}{I_{te}} \\[2mm] \sigma_{cc} = \sigma_c + \sigma_{cc}' = \dfrac{E_c}{E_s}\left(\dfrac{M_x y}{I_e} + \dfrac{M_t x}{I_{te}}\right) \\[2mm] \sigma_{st} = \sigma_s + \sigma_{st}' = \dfrac{M_x y}{I_e} + \dfrac{M_t x}{I_{ts2}} \end{cases} \tag{6.16}$$

where σ_{sc}, σ_{cc}, and σ_{st} are the normal stress in the top steel tube, in the concrete, and in the bottom steel tube, respectively.

The above theoretical analysis is based on the configuration of an I-girder with CFTF and CW, as shown in Fig. 6.5. Such I-girder is generally used in a frame structure with a fixed-fixed constraint at its both ends. When the I-girder is used in a pinned-pinned boundary support, such as in bridge, the bottom flange is generally a flat steel plate rather than a CFTF. For this type of I-girders, the normal stress analysis is similar as the above method except replacing the geometric parameters of CFTF (such as moment of inertia, height, thickness, area, and so on) with corresponding results of a flat plate.

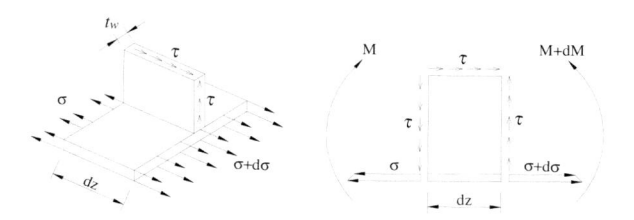

A shear stress in the straight segment of the corrugated web

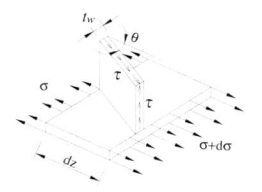

B shear stress in the inclined segment of the corrugated web

Figure 6.8 Shear stress in the corrugated web.

6.3 Shear stress in corrugated web

To present equation for calculating the shear stress on the CW, an infinitesimal segment of the new I-girder in the length direction is taken for analysis, and the shear stress of a location at the CW with a distance of y to the bottom flange as shown in Fig. 6.8A and B. When the location is located at the straight segment of the trapezoidal web, as shown in Fig. 6.8A, the force equilibrium in the length direction of the new I-girder produces the following equation:

$$\tau \cdot t_w \cdot dz = \int_{A_{f2}} \frac{dM_x}{I_e} y dA \tag{6.17}$$

where A_{f2} is the area of the bottom flange. The shear stress is then calculated from the following equation:

$$\tau = \frac{1}{t_w I_e} \int_{A_{f2}} \frac{dM_x}{dz} y dA = \frac{V_y}{t_w I_e} \int_{A_{f2}} y dA \approx \frac{V_y A_{f2} y_n}{t_w I_e} \tag{6.18}$$

where, t_w is the thickness of the CW.
  If an equivalent area A_{we} is denoted as follows:

$$A_{we} = \frac{t_w I_e}{A_{f2} y_n} \tag{6.19}$$

Then, Eq. (6.18) can be expressed as follows:

$$\tau = \frac{V_y}{A_{we}} \tag{6.20}$$

As it is easy to see from Eq. (6.19) that A_{we} is a constant for a given new I-girder, the shear stress on the cross-section in the straight segment of the CW is uniform.

The shear stress in the inclined segment of the CW can be also calculated from the similar method. As seen from Fig. 6.8B, the force equilibrium in the length direction is expressed as follows:

$$\tau \cdot A \cdot \cos\theta = \tau \cdot t_w \cdot \frac{dz}{\cos\theta} \cdot \cos\theta = \tau \cdot t_w \cdot dz = \int_{A_{f2}} \frac{dM_x}{I_e} y \, dA \tag{6.21}$$

Obviously, the force equilibrium of the inclined segment in length direction of the new I-girder is identical to Eq. (6.17). Therefore, Eq. (6.18) is also applicable for the inclined segment of the CW. In overall, Eq. (6.28) can be used for calculating the shear stress along the height of the CW in the new I-girder.

The shear force sustained by the CW is calculated from the following equation:

$$V_y^{cor} = \tau \cdot A_w = \frac{V_y}{A_{we}} \cdot t_w \cdot (H - h_u - t_f) \approx \frac{V_y A_{f2} y_n (H - h_u)}{I_e} \tag{6.22}$$

where A_w is the the area of the cross section of the CW. Therefore, the percentage of shear forced sustained by the CW is estimated from the following equation:

$$\eta = \frac{V_y^{cor}}{V_y} = \frac{A_{f2} y_n (H - h_u)}{I_e} \tag{6.23}$$

When η is not very close to 1.0, the top CFTF also sustains some amount of shear force in the new I-girder, which indicates that it is not suitable to assume that the CW sustains most of the shear force.

6.4 Flexural yielding strength of I-girders

In the above session, equations for calculating the stress distribution on the cross-section of I-girders with CFTF and CW are presented. From such equations, yielding initiation at the top and bottom flanges can be determined by setting the maximum normal stress of the steel tube to be the yield stress of the steel materials. The corresponding bending moment (M_e) can be also calculated at this state. However, the flexural yielding strength of the girder is much bigger than M_e because a steel tube allows plasticity to develop continuously along its height direction. When plasticity distributes fully on the cross-section of the bottom steel tube, as illustrated in Fig. 6.9. Accordingly, the stresses on the top surface and on the bottom surface of the top flange

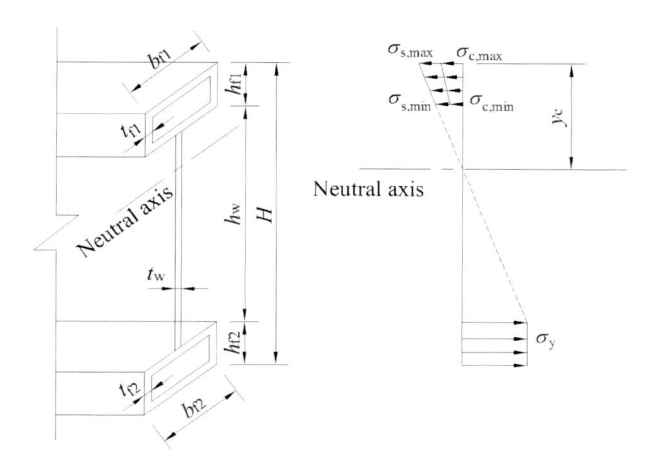

Figure 6.9 Normal stress distribution in flanges at yielding strength.

steel tube may be still in an elastic state due to the contribution of the infilled concrete, and they are denoted with $\sigma_{s,max}$ and $\sigma_{s,min}$, respectively. The corresponding values of the infilled concrete are $\sigma_{c,max}$ and $\sigma_{c,min}$, respectively. In Fig. 6.9, y_c is the distance of the neutral axis to the top surface of the top flange.

$\sigma_{s,max}$ and $\sigma_{s,min}$ can be obtained through deduction as expressed in the following equations:

$$\sigma_{s,max} = \frac{y_c}{h_w + h_{f1} - y_c} \cdot \sigma_y \tag{6.24}$$

$$\sigma_{s,min} = \frac{y_c - h_{f1}}{h_w + h_{f1} - y_c} \cdot \sigma_y \tag{6.25}$$

The average stress $\overline{\sigma_{sc}}$ of the top steel tube is then calculated from the following equation:

$$\overline{\sigma_{sc}} = \frac{y_c - h_{f1}}{h_w + h_{f1} - y_c} \cdot \sigma_y \tag{6.26}$$

Based on the assumption that no relative slip occurs at the contact surface of concrete and steel tube, identical normal strains exist between steel tube and infilled concrete at the same cross-section, i.e., $\frac{\sigma_s}{E_s} = \frac{\sigma_c}{E_c}$. Therefore, the relationship of stresses in the infilled concrete (σ_c) and in the top steel tube (σ_s) can be expressed as follows:

$$\sigma_c = \frac{E_c}{E_s} \cdot \sigma_s \tag{6.27}$$

The average stress $\overline{\sigma_{cc}}$ of the infilled concrete in the top steel tube is then calculated from the following equation:

$$\overline{\sigma_{cc}} = \frac{\sigma_{c,\max} + \sigma_{c,\min}}{2} = \frac{\frac{E_c}{E_s} \cdot \sigma_{s,\max} + \frac{E_c}{E_s} \cdot \sigma_{s,\min}}{2} = \frac{E_c}{E_s} \cdot \overline{\sigma_{sc}} \tag{6.28}$$

From the equilibrium equation of the axial force on the cross-section, i.e., $\Sigma F_x = 0$, the following equation is satisfied:

$$\sigma_y \cdot A_{st} = \overline{\sigma_{sc}} \cdot A_{sc} + \overline{\sigma_{cc}} \cdot A_{cc} \tag{6.29}$$

The location of the neutral axial (y_c) can be determined by substituting Eqs. (6.26) and (6.28) into Eq. (6.29), and it is listed as follows:

$$y_c = \frac{\frac{1}{2} \cdot h_{f1} + \eta_1 \cdot \left(h_w + h_{f1} \right)}{1 + \eta_1} \tag{6.30}$$

Once y_c is determined, the flexural yielding strength, in terms of plastic bending moment (M_p), can be obtained easily from the following equations:

$$M_p = M_{st} + M_{sc} + M_{cc} = \sigma_y \cdot A_{st} \cdot \left(H - y_c - \frac{h_f 2}{2} \right) + \overline{\sigma}_{sc} \cdot A_{sc} \cdot \left(y_c - \frac{h_{f1}}{2} \right)$$
$$+ \overline{\sigma}_{cc} \cdot A_{cc} \cdot \left(y_c - \frac{h_f 1}{2} \right)$$

$$\tag{6.31}$$

where M_{st}, M_{sc}, and M_{cc} are the components of bending moment for the bottom steel tube, for the top steel tube and for the infilled concrete in the top flange respectively.

When the bottom flange is a flat steel plate, plasticity in it can be ignored because the thickness of the steel plate is much smaller compared to the height of the cross-section. In this case, the flexural yielding strength can be predicted from Eq. (6.16) by setting the normal stress in the bottom flange to be the yield stress of the steel materials. The detailed calculation is not introduced here.

6.5 Experimental verification

6.5.1 Specimens and test setup

To verify the accuracy of the presented equations for calculating the normal and the shear stress in the new I-girder, an efficient method is used to compare the predicted normal and shear stresses from the presented equations with experimental results obtained from tests. A total of four I-girder specimens are designed and tested. In the four specimens, three ones (SP1-1~SP1-3) are designed with top CFTF and bottom flat plate flange, as shown in Fig. 6.10A, while the fourth one (SP2) is designed with both top and bottom CFTFs, as shown in Fig. 6.10B. The tested specimens are shown in

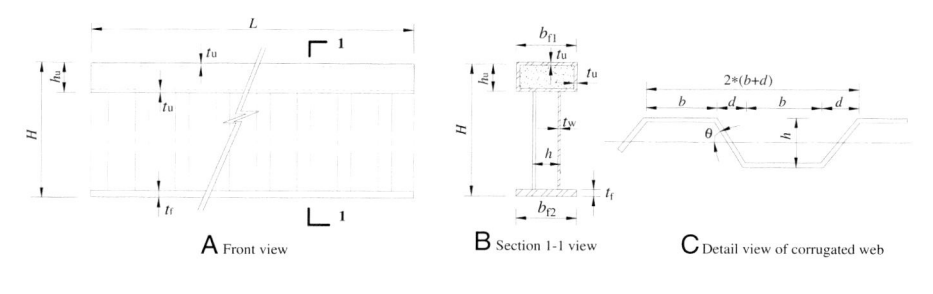

A Top CFTF and bottom flat plate flange

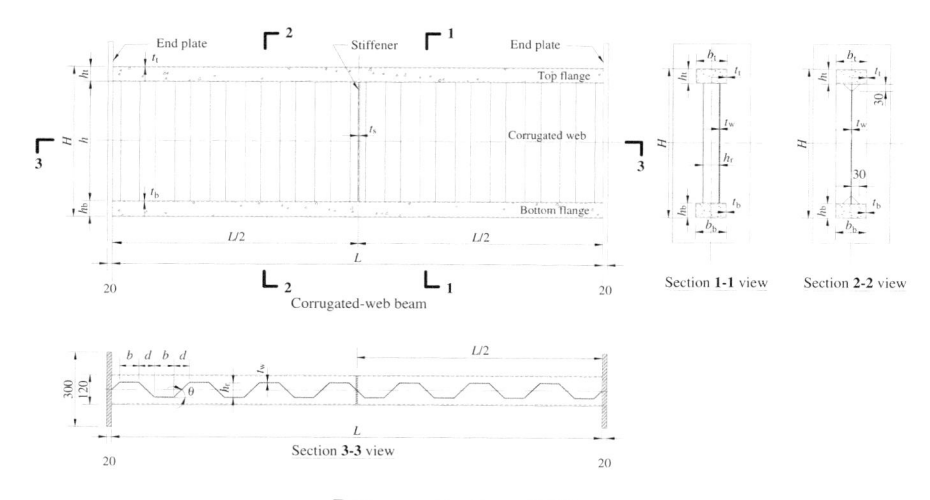

B Top and bottom CFTFs

Figure 6.10 Geometry of I-girder specimens.

Table 6.1 Geometric dimensions of I-girder specimens.

Specimen	L [mm]	H [mm]	t_f [mm]	b_f [mm]	t_w [mm]	t_u [mm]	h_u [mm]
SP1-1	6000	600	8	160	3	3	80
SP1-2	2000	600	6	160	3	3	80
SP1-3	2186	600	5	120	3	3	60
SP2	1862	600	3	120	3	3	60

Fig. 6.11. The dimensions of the four specimens, including the length (L) and the height (H) of the girder, the width (b_f) and the height (h_u) of the steel tube of the CFTF, the thickness of the steel tube (t_u), the thickness of the flat plate flange (t_f), and the thickness of the web (t_w), are all tabulated in Table 6.1. The dimensions of the CWs are given in Table 6.2. The mechanical properties of the steel materials, measured from tensile tests on coupons, are listed in Table 6.3. For the infilled concrete, the characteristic

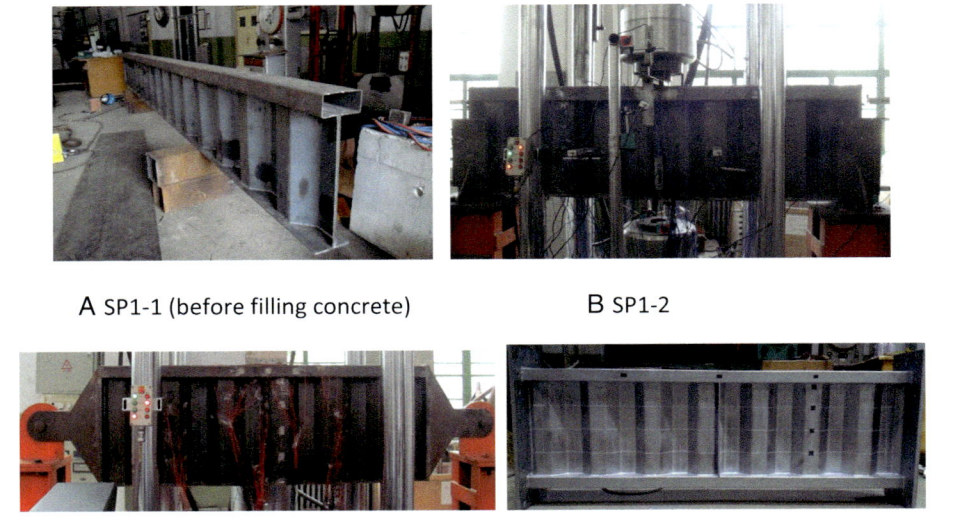

A SP1-1 (before filling concrete)   B SP1-2

C SP1-3   D SP2

Figure 6.11 View of four I-girder specimens.

Table 6.2 Geometric dimensions of trapezoidal web.

Specimen	b [mm]	d [mm]	h [mm]	θ [°]
SP1-1	110	90	90	45
SP1-2				
SP1-3				
SP2	73	60	60	45

Table 6.3 Mechanical properties of steel materials.

	Yielding strength [MPa]	Tensile strength [MPa]	Elastic modulus [GPa]	Elongation
Corrugated web	341	473	208	21%
Flat-plate flange	318	481	201	22%
Steel tube	350	419	208	13.6%

compressive strength is 30.2 MPa and the elastic modulus is 30 GPa.

All four specimens are tested through the three-point loading method, as illustrated in Fig. 6.12. The specimens are supported at both ends without restricting the in-plane rotation to simulate a pinned-pinned connection. A concentrated load is applied at the mid-span. During the loading process, normal strains at some specified locations of the flanges and the CW are measured from placed strain gauges. For example, two selected locations, namely P_1 and P_2, are specified in specimens SP1-1 and SP1-3, as shown in Fig. 6.13. Strain gauges are placed in longitudinal direction in the two locations to

A SP1-1

B SP1-2 **C** SP1-3 **D** SP2

Figure 6.12 Testing setup.

measure the normal strains. In addition, strain gauges in longitudinal direction, as strain gauges 1~5 in Fig. 6.13B, are also placed along the web height to measure the normal strain distribution in the CW to verify the so-called accordion effect. Three groups of strain rosettes, as shown in Fig. 6.13B, are placed in three locations (Location-1, Location-2, and Location-3) of SP1-3 along the web height. From the normal strains measured from the strain rosettes in three directions (0°, 45°, and 90°) can be used to calculate the shear strain. Similarly, five strain rosettes are also placed in SP2, as shown in Fig. 6.13C, to calculate the shear strain distribution along the web height.

To measure both in plane and out-of-plane deformations of the specimens, linear variable displacement transducers (LVDTs) are placed at some critical positions during the static tests. For example, for specimen SP2, the arrangement of the LVDTs is illustrated in Fig. 6.14. Five LVDTs (LVDT-1~LVDT-5) are placed at the bottom CFTF to measure the lateral displacement which is produced by the out-of-plane moment induced by shear stress flow in the CW. LVDTs 6~7 are placed at the mid-height of the CW to monitor if local buckling occurs during the loading process. LVDTs 8~9 are placed at two locations of the endplate corresponding to the top and bottom flanges, and they are used to measure the rotation of the specimen under bending moment. LVDT 10 is placed at the mid-span of the bottom flange to measure the in-plane deflection of the specimen. LVDT 11 is placed at the top flange to monitor if global buckling occurs.

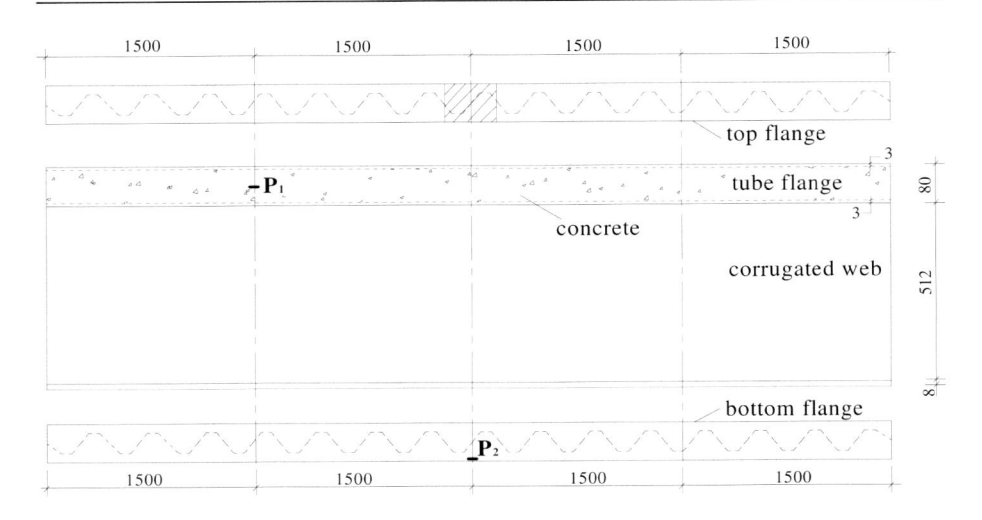

A SP1-1

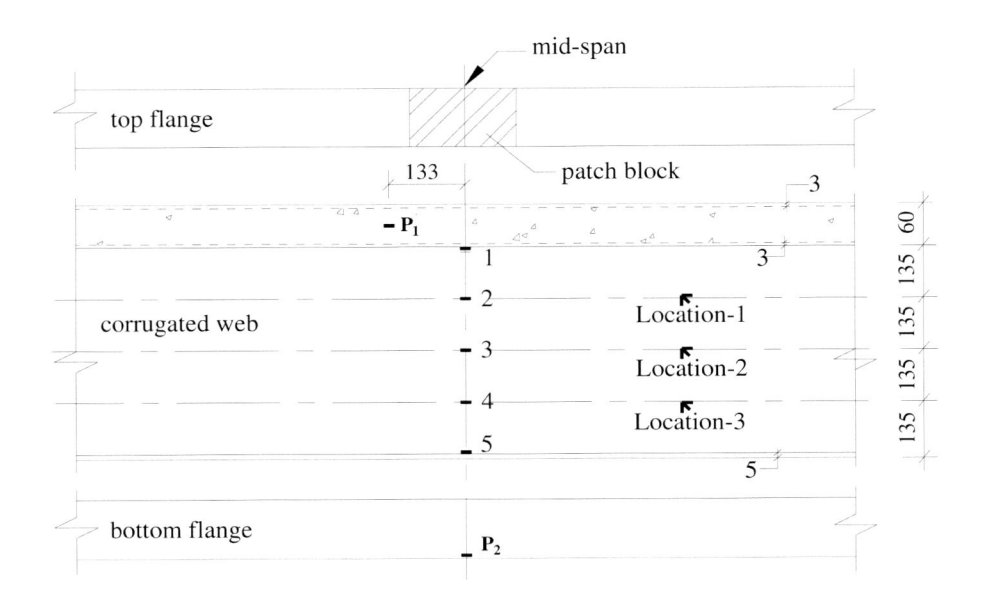

B SP1-3

Figure 6.13 Placement of strain gauges in specimens SP1-1, SP1-3, and SP2.

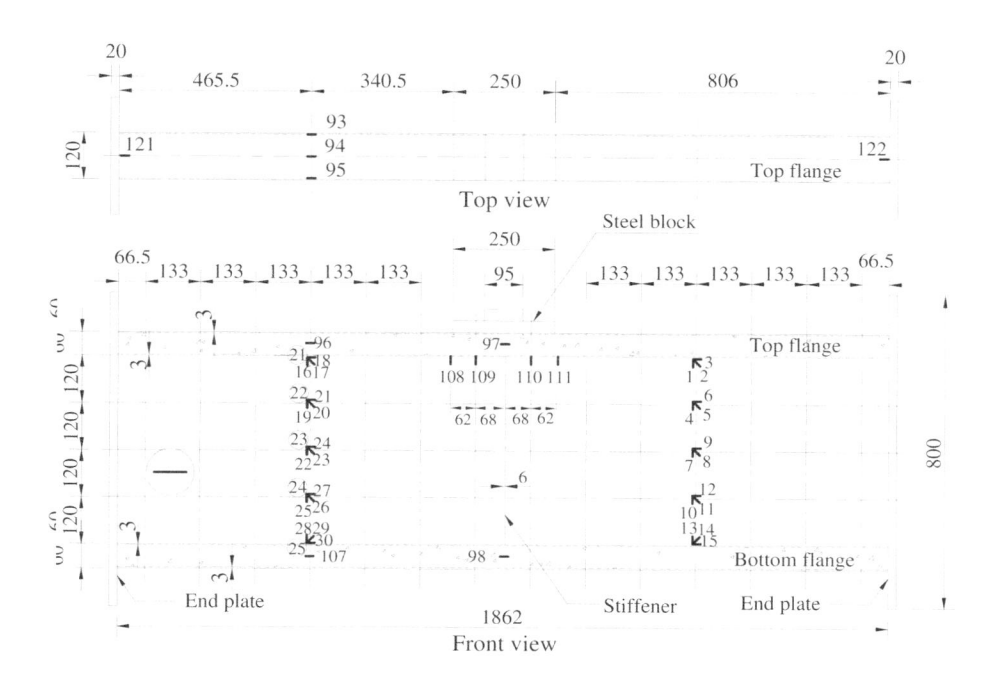

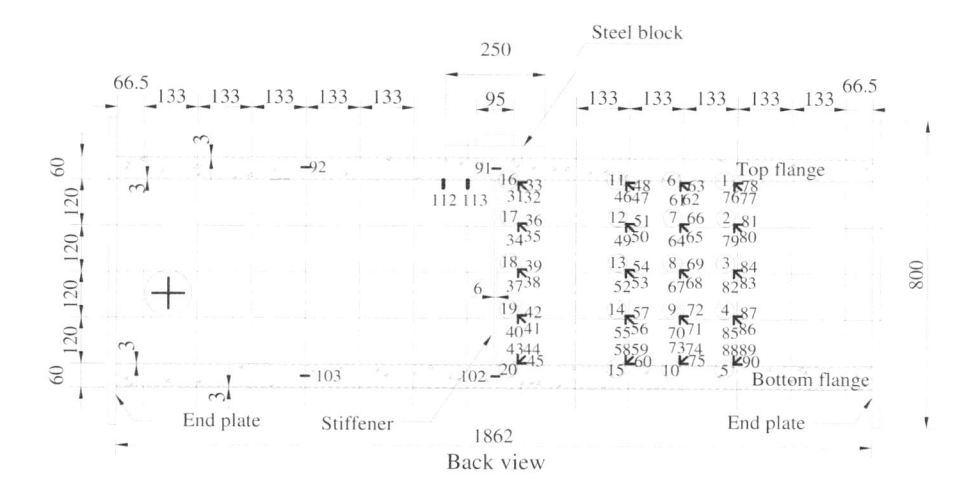

C SP2

Figure 6.13, cont'd.

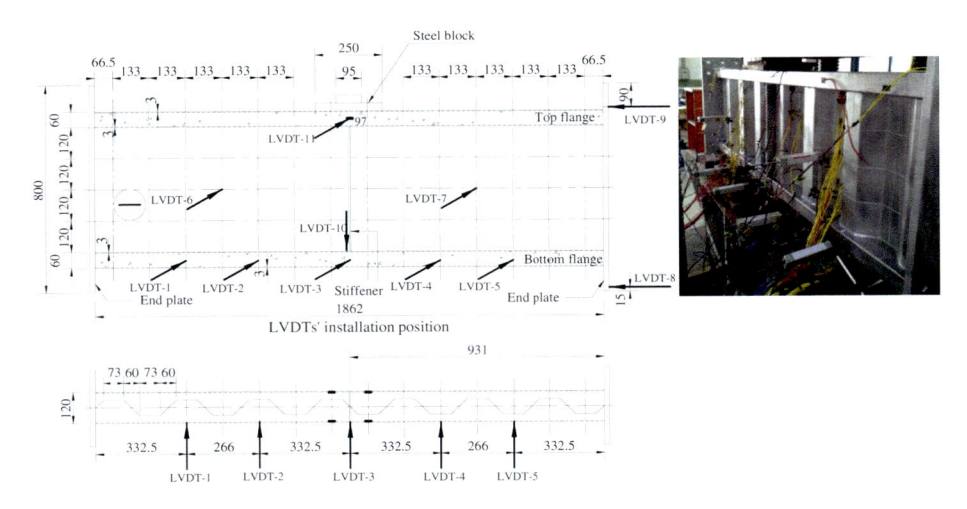

Figure 6.14 Placement of LVDTs in specimen SP2. *LVDTs*, linear variable displacement transducers.

6.5.2   Results and discussions

The failure modes of the four specimens are shown in Fig. 6.15, in which the yielding of the bottom flat plate flange is the most important mechanism for the load-carrying capacity. The global flexural deformation is the dominant failure mode for the specimens with bottom flat plate flange, as shown in Fig. 6.15A–C. Because a concrete-filled steel tube has much bigger stiffness compared to a flat steel plate, specimen SP2 with bottom CFTF has much smaller flexural deformation compared to the other three specimens, as shown in Fig. 6.15D.

As mentioned previously, due to the shear stress flow in the CW, the flanges are subjected to out-of-plane bending moment, as illustrated in Fig. 6.16, and such bending moment produces wave-shape out-of-plane flexural deformation to the flanges. Because the CFTF has much bigger flexural stiffness compared to the flat plate flange, the wave-shape deformation of the top CFTF is very small (as shown in Fig. 6.16A) while such deformation is very clear, as shown in Fig. 6.16B. It is noted that global buckling is not observed in all specimens although the bottom flat plate flange has very clear residual deformation, which indicates that the CFTF improves the resistance of the girders to lateral displacement efficiently.

The normal stress distribution along the web height is illustrated in Fig. 6.17. Fig. 6.17A shows the normal stresses at five locations along with the web height in specimen SP1-3, as shown in Fig. 6.11C. Although the normal stresses at the five locations increase with the increase of the load, the values are very small and much lower than the yield strength even at the ultimate strength of the specimen. Fig. 6.17B shows the normal strain distribution along the web height in specimen SP2. Except the normal strains at the flange/web connection, the normal strains are also very small at different loading levels. The experimental results in Fig. 6.17 verify the correctness of

A SP1-1   B SP1-2

C SP1-3   D SP2

Figure 6.15 Failure mode.

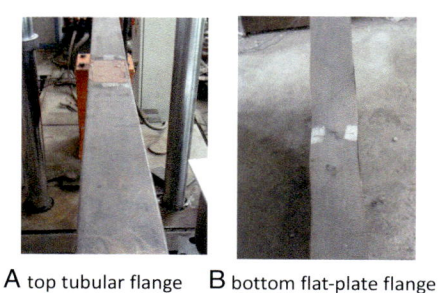

A top tubular flange   **B** bottom flat-plate flange

Figure 6.16 Deformation of flanges in SP3.

the assumption that the CW does not sustain the tension and compression produced by the bending moment.

As predicted from Eqs. 6.19 and 6.20, the shear strain/stress distribution along the web height is uniform. This conclusion is verified from experimental results, as shown in Fig. 6.18. In Fig. 6.18A, the shear stresses of specimen SP1-3 at three different locations along with the web height (as seen in Fig. 6.18B) are almost the same at different loading levels. Similarly, the shear strain distributions of specimen SP2 along the web height are also very close to the same values. For both specimens, predictions of shear stress or shear strain from Eq. (6.20) agree very well with experimental measurements. The experimental results verify the uniform distribution of the shear

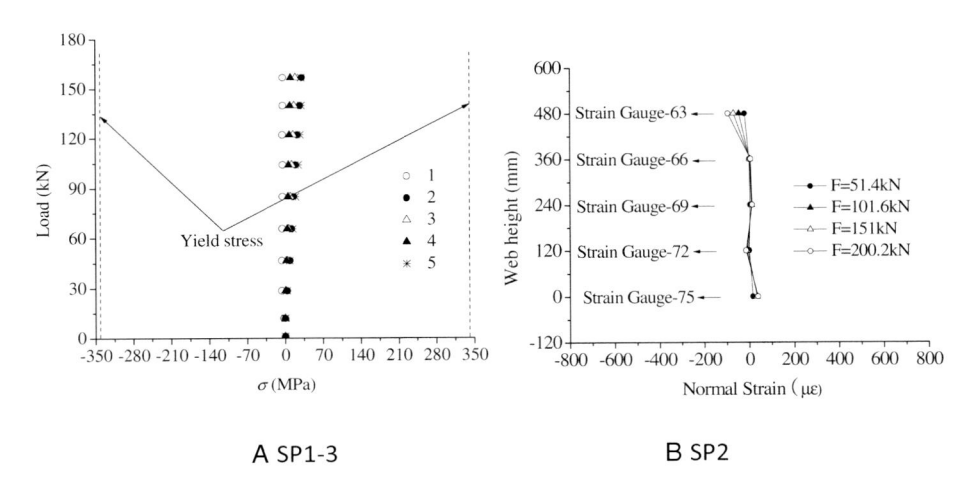

A SP1-3

B SP2

Figure 6.17 Normal stress development on the web.

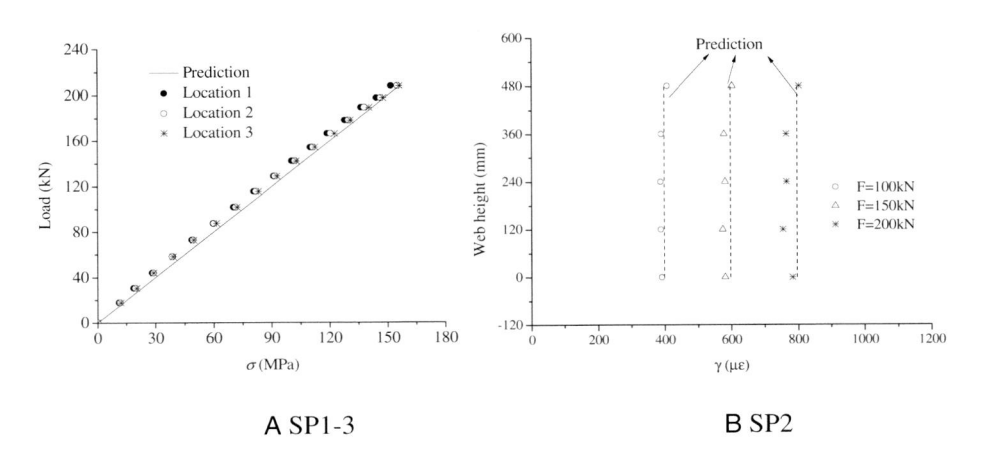

A SP1-3

B SP2

Figure 6.18 Shear stress development on the web.

stress along the height in the CW, and the presented equation is also verified to be reliable to predict the shear stress in the CW.

The normal stress development in the flanges can be predicted from Eq. (6.16), and the predictions can be also verified through comparison with experimental results. In Fig. 6.19A and B, the normal stresses at two selected positions on top and bottom flanges of specimen SP1-1 (P1 and P2, shown in Fig. 6.11A) are calculated from Eq. (6.16), and the predictions are plotted together with test results. Generally, the predictions of the normal stress in the bottom flat plate flange agree well with test results, as shown in Fig. 6.19B. However, the predictions of the normal stress in the top CFTF seem to be smaller compared to test results, as shown in Fig. 6.19A. The reasons can be generalized as follows: (1) the bottom flat plate has a bigger thickness compared

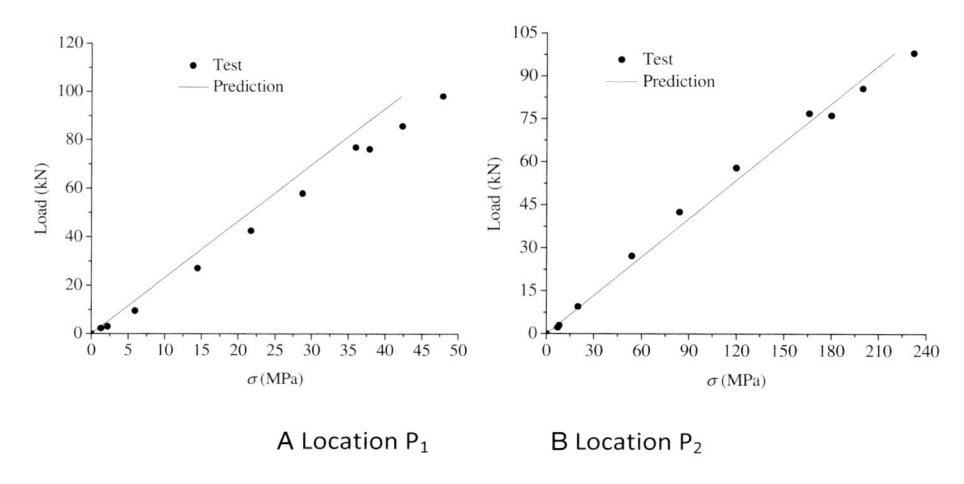

A Location P_1 B Location P_2

Figure 6.19 Comparison of stress development in locations P_1 and P_2 for SP1-1.

to the bottom steel tube, and hence the geometric imperfection of the bottom flange is smaller. Imperfection produces less effect on the normal stress in the bottom flange. (2) The top steel tube is subjected to compression, and a thin-walled steel tube under compression is much more sensitive to imperfection. (3) Due to the collaboration of the infilled concrete, the resistance of the top CFTF to compression is improved efficiently, and the compressive normal stress in the steel tube is much smaller than the tensile normal stress in the bottom flat plate flange, as shown in Fig. 6.19A and B. Obviously, a much smaller stress level will produce larger measuring error in experiment. Based on the above three reasons, the prediction of the normal stress in top steel tube is not as accurate as the prediction in the bottom flat plate. Nevertheless, the predictions of the normal stresses in both top and bottom flanges are acceptable compared to experimental results, and Eq. (6.16) can be used to calculate the normal stress in the flanges of the girders.

Because the normal stress in the flange can be predicted from Eq. (6.16), the yielding strength of the specimens can be easily obtained for specimens SP1-1~SP1-3 by setting the normal stress in the bottom flat plate flange to be the yield strength of the steel material, as shown in Figs. 6.20A–C. The predictions of the yielding strength of the three specimens with bottom flat steel plate seem to be very similar with the test results, which can be evaluated from the boundary between linear and nonlinear stages in the load-deflection curve. For specimen SP2, the prediction on the yielding strength is different because yielding region in the bottom steel tube can develop in the height direction till the whole cross-section of the bottom steel tube is in yielding state. In this case, the yielding strength of specimen SP2 is predicted from Eq. 6.31. As seen from Fig. 6.20D, prediction of the yielding strength of SP2 also agrees very well with experimental result. Eqs. (6.16) and (6.31) are both accurate in calculating the yielding strength of the girders with bottom flat plate flange or CFTF.

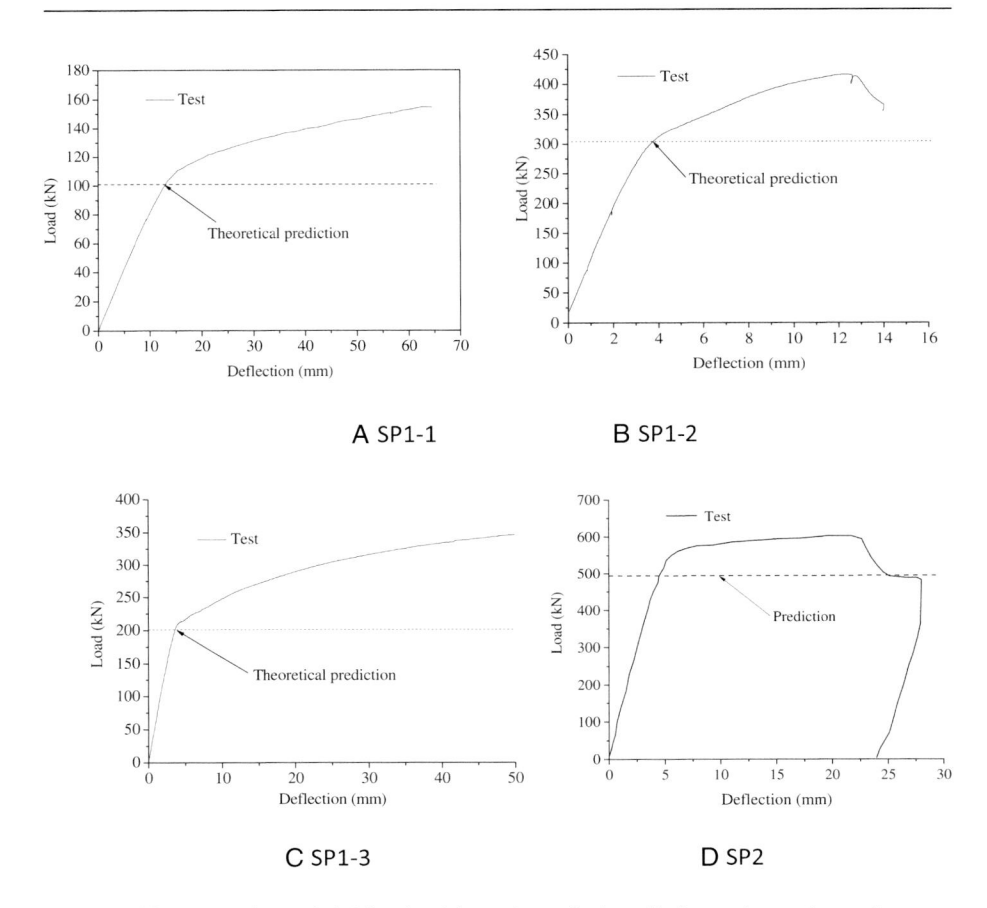

A SP1-1   **B** SP1-2

C SP1-3   **D** SP2

Figure 6.20 Comparison of yielding load from theoretical prediction and experimental test.

6.6   Numerical verification

In the experimental verification on the presented analytical method, only one specimen (SP2) is tested experimentally. To provide convincing assessment on the accuracy and reliability of the presented analytical method, numerical simulation can be also used to carry out verification. In the numerical analysis, software ABAQUS is used to build finite element (FE) models and carry out analysis for the failure process of specimen SP2. By comparing numerical results with experimental results, the verified FE model can be then used to conduct parametric study on more different models for further comparison with predictions.

6.6.1   Finite element modeling

For mechanical properties of steel, the experimental data obtained from the coupon tests are used in the FE simulation, while the σ-ε constitutive relation of concrete

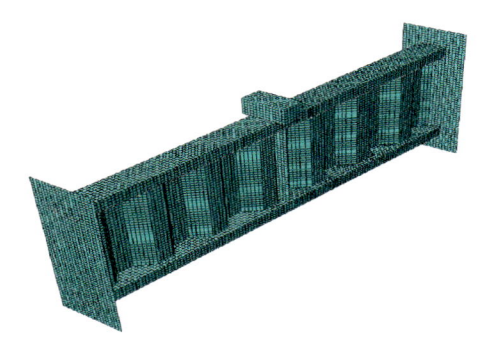

Figure 6.21 FE mesh. *FE*, finite element.

proposed by Carreira et al. [16] and Hassan [17] are utilized to represent the compressive behavior and tensile behavior respectively. Poisson's ratios for steel and concrete are 0.3 and 0.2, respectively, which have been used widely in FE simulations in literature.

In the mesh generation, 8-node brick elements with three translation degrees of freedom at each node (C3D8R) are used to discretize the infilled concrete, while 4-node shell elements with reduced integration (S4R) are chosen to discretize the steel tube and the steel plate in the flanges and in the web respectively. The endplates of the specimens are considered as a rigid body with the element type of S4R. Mesh convergence studies are carried out to set optimal mesh and a 20 mm mesh size is finally used for the whole model, as shown in Fig. 6.21, which could provide solution with relatively high accuracy and low computational time. For a typical specimen, it contains over 10000 elements in total. Thus, the reliability of the simulation results is ensured.

In the current FE model, "Merge" command in the ABAQUS is used to simulate the welding effect, while a "Surface-to-Surface Contact" is set between the steel tube and inner filled concrete. For the interface connection, "Hard Contact" is created in the normal direction, which allows the separation of the interface in tension and no penetration in compression. Coulomb friction model is used in tangent contact with the friction coefficient of 0.6, as recommended by Han et al. [18], which agrees with most test results conducted by Rabbat et al. [19]. Finally, a "Tie" constraint is defined between top loading plate (rigid steel block) and the top flange surface at the mid-span.

For the boundary conditions in the FE model, a coupling constraint is established between each of the endplates and a reference point. As a result, the degree of freedoms (DOFs) of the endplate is associated with that of the reference point. The boundary conditions are consistent with the actual situation in experiment. The endplate is placed on a support and thus a hinged connection is used here. In this case, the reference point is fixed for all DOFs except the in-plane rotation to achieve the hinged boundary condition. In addition, displacement control method is adopted and applied to the loading steel plate on the top surface of the top CFTF at the mid-span to carry out FE analysis.

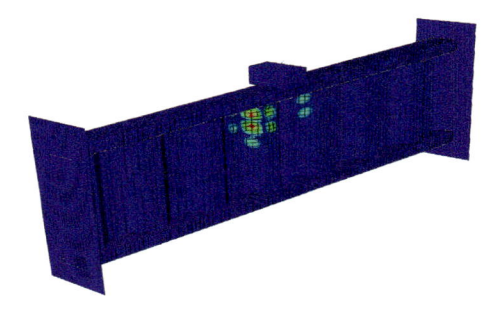

Figure 6.22 First buckling mode from FE simulation. *FE*, finite element.

In the FE model, initial imperfections are also considered. A linear elastic buckling analysis is conducted to obtain the first buckling mode, as shown in Fig. 6.22, and then the buckling shapes are scaled and imported into the perfect geometry of the girders to simulate the initial imperfection. As illustrated in Fig. 6.22, the first mode buckling occurs on the web for both specimens, which indicates that the initial imperfection of out-of-plane deformation for the steel tubes on the top and bottom flanges plays a minor role and could be ignored in the finite element model. Such idea was confirmed in previous research by Tao et al. [20]. Therefore, only web buckling is considered as the initial geometric imperfection in current FE simulation.

6.6.2 Verification of finite element models

To verify the reliability of the FE model, the failure mode and the load–displacement curve obtained from the FE analysis are compared with those from the experimental test. To present clearly the failure process of the tested specimen SP2, the stress development is shown in Fig. 6.23. In the early stage (especially the elastic stage), the stress on the whole web remains the same along with the height, as shown in Figs 6.23A and B. As it is found from experimental results that tensile stress is negligible in the CW, such uniform stress refers mainly to the shear stress component, which verifies the conclusion from experimental results that the shear stress on the CW is uniformly distributed along the height. As the load increases gradually, both the top and bottom flanges of the specimen start to yield, and the yielding area extends continuously along the longitudinal direction (Fig. 6.23C and D). Finally, the vertical deformation of the girder at the mid-span increases promptly due to the yielding on the cross-section of both top and bottom CFTFs at the mid-span, as seen from Figs 6.23E and F. The failure mechanism of SP2 assumes the flexural yielding of the CFTFs. In overall, the failure mode obtained from FE simulation agrees well with the experimental observation, and the comparison between the two results is shown in Fig. 6.24. The comparison indicates that the FE model is able to simulate the failure process of I-girders with CFTFs and CW accurately.

The load–displacement curve at the mid-pan obtained from FE analysis is also compared with experimental result to verify the accuracy of the FE model, and

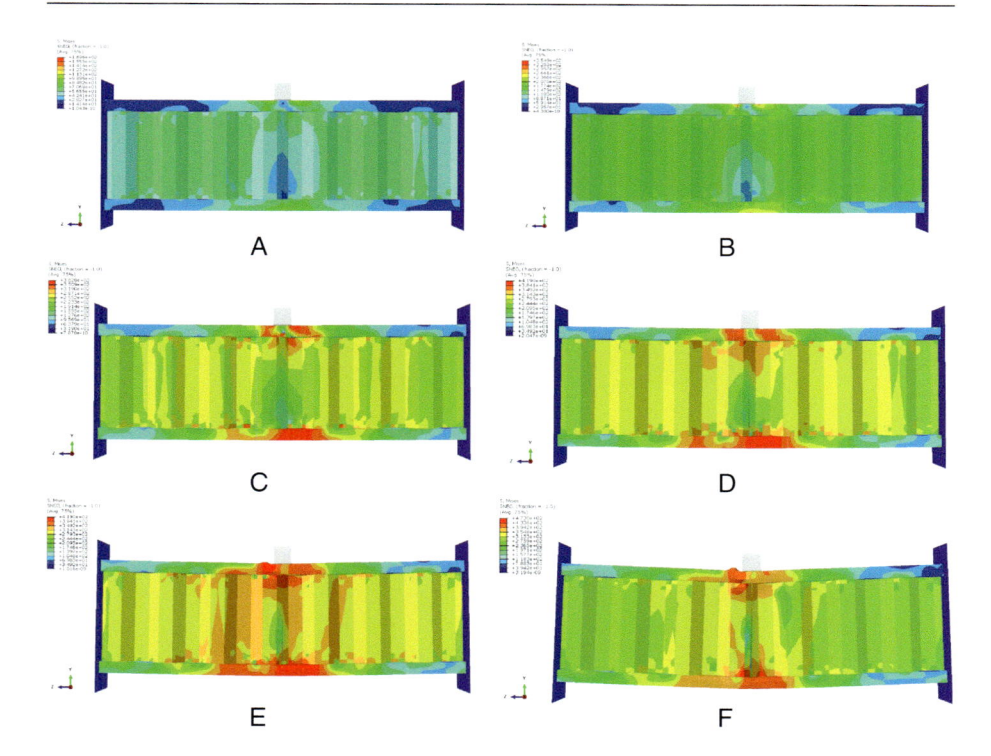

Figure 6.23 Stress development of SP2.

the comparison is shown in Fig. 6.25. The FE simulation of the load–displacement curve agrees quite well with experimentally measured result, and the FE model is verified again from such comparison. Hence, the presented FE model can be used reliably to carry out more numerical analyses for the composite I-girders with different parameters.

6.6.3 Parametric study

To carry out further verification on the accuracy of the presented equation for predicting the yield strength (Eq. 6.31), a parametric study by using FE simulation is conducted. Since the FE model has been evaluated to be accurate enough, the FE simulations can be also used to verify the accuracy of the presented equation. In the parametric study, both geometric parameters (h_w, h_f, t_f, b_f) and material property factors (yield strength of steel material σ_y, characteristic compressive strength of concrete σ_c) are considered. The FE models in the parametric study consist of three groups: (1) web height h_w; (2) geometric configuration of steel tube flange, such as the flange height h_f (including h_t and h_b, in this section, $h_t = h_b = h_f$), the thickness of flange tube t_f ($t_t = t_b = t_f$); and the width of flange tube b_f ($b_t = b_b = b_f$); (3) material properties including σ_y and σ_c. A total of 30 FE models are analyzed in the parametric study and the detailed information

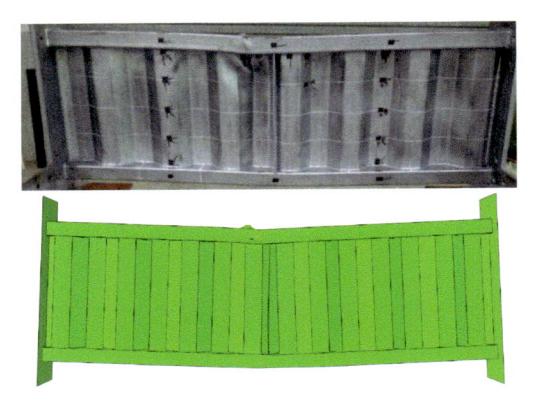

Figure 6.24 Comparison of failure mode.

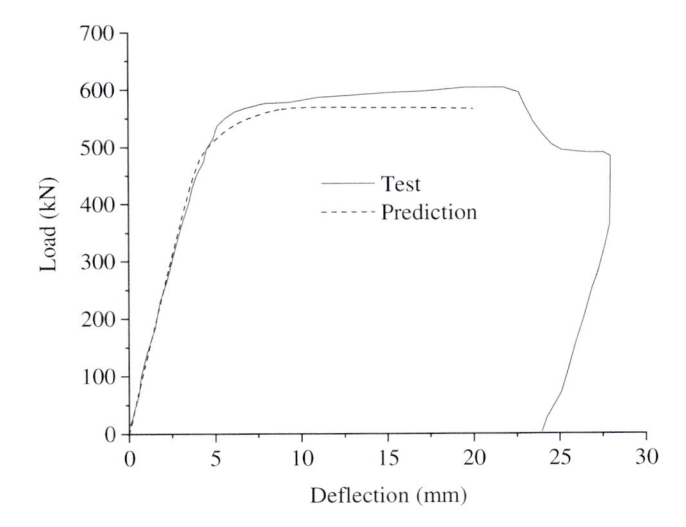

Figure 6.25 Comparison of load–displacement curve.

on the FE models is listed in Table 6.4. In the parametric study, the length of the girder (L), the thickness of the web (t_w), and the elastic modulus of steel material (E_s) are set with the values of 1862 mm, 5 mm, and 200 GPa, respectively. In addition, hinged boundary condition is used at the endplate.

The comparison between the predicted flexural strengths (N_p) and the FE simulations (N_{FE}) for the 30 models is listed in Table 6.4 and plotted in Fig. 6.26. In summary, the mean value (Mean) and standard deviation (STD) of the ratio N_{FE}/N_p are 1.06 and 0.0235, respectively. Such small error shows a good agreement between the predicted and the FE results. As observed from Fig. 6.26 that the difference of the two results

Table 6.4 Information of finite element models in parametric study.

Model	h_w [mm]	$h_f \times t_f \times b_f$ [mm]	σ_y [MPa]	σ_c [MPa]	N_{FE} [kN]	N_P [kN]	N_{FE}/N_p [%]	Mean	STD
FE1	400	$60 \times 3 \times 120$	235	30	303.6	280.0	1.08	1.08	0.032
FE2	480	$60 \times 3 \times 120$	235	30	356.2	328.7	1.08		
FE3	560	$60 \times 3 \times 120$	235	30	409.0	377.5	1.08		
FE4	640	$60 \times 3 \times 120$	235	30	459.8	426.2	1.08		
FE5	720	$60 \times 2 \times 120$	235	30	510.5	474.9	1.07		
FE6	480	$40 \times 3 \times 120$	235	30	304.2	280.2	1.09	1.088	0.057
FE7	480	$50 \times 3 \times 120$	235	30	328.8	304.1	1.08		
FE8	480	$60 \times 3 \times 120$	235	30	356.2	328.7	1.08		
FE9	480	$70 \times 3 \times 120$	235	30	386.2	354.1	1.09		
FE10	480	$80 \times 3 \times 120$	235	30	417.6	380.1	1.10		
FE11	480	$60 \times 2 \times 120$	235	30	247.5	221.7	1.12	1.07	0.0221
FE12	480	$60 \times 3 \times 120$	235	30	356.2	328.7	1.07		
FE13	480	$60 \times 4 \times 120$	235	30	458.4	433.3	1.05		
FE14	480	$60 \times 5 \times 120$	235	30	562.7	535.3	1.05		
FE15	480	$60 \times 6 \times 120$	235	30	668.5	634.8	1.05		
FE16	480	$60 \times 3 \times 90$	235	30	298.0	272.1	1.10	1.07	0.0153
FE17	480	$60 \times 3 \times 120$	235	30	356.2	328.7	1.08		
FE18	480	$60 \times 3 \times 150$	235	30	409.8	385.4	1.06		
FE19	480	$60 \times 3 \times 180$	235	30	466.4	442.1	1.05		
FE20	480	$60 \times 3 \times 210$	235	30	524.8	498.8	1.04		
FE21	480	$60 \times 3 \times 120$	235	50	355.3	328.7	1.08	1.046	0.0198
FE22	480	$60 \times 3 \times 120$	295	50	437.1	412.7	1.06		
FE23	480	$60 \times 3 \times 120$	345	50	502.8	482.6	1.04		
FE24	480	$60 \times 3 \times 120$	390	50	560.6	545.6	1.03		
FE25	480	$60 \times 3 \times 120$	420	50	599.9	587.5	1.02		
FE26	480	$60 \times 3 \times 120$	345	40	501.3	482.6	1.04	1.037	0.028
FE27	480	$60 \times 3 \times 120$	345	50	501.2	482.6	1.04		
FE28	480	$60 \times 3 \times 120$	345	60	497.9	482.6	1.03		
FE29	480	$60 \times 3 \times 120$	345	70	501.5	482.6	1.04		
FE30	480	$60 \times 3 \times 120$	345	80	502.1	482.6	1.04		

Mean value: 1.06; Standard deviation (STD): 0.024.

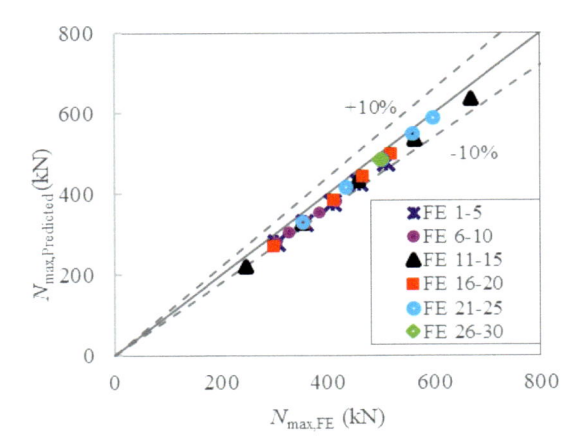

Figure 6.26 Comparison of flexral strengths between FE simulations and predictions. *FE,* finite element.

is all within 10%. Therefore, the presented equations are reliable in calculating the flexural strength of an I-girder with CFTFs and CW.

References

[1] Y.L. Pi, N.S. Trahair, Lateral-distortional buckling of hollow flange beams, J Struct. Eng., ASCE 123 (1997) 695–702.

[2] B. Kim, R. Sause, Lateral torsional buckling strength of tubular flange girders, J Struct. Eng., ASCE 134 (2008) 902–910.

[3] T. Anapayan, M. Mahendran, D. Mahaarachchi, Lateral distortional buckling tests of a new hollow flange channel beam, Thin Wall. Struct. 49 (2011) 13–25.

[4] T. Anapayan, M. Mahendran, Improved design rules for hollow flange sections subject to lateral distortional buckling, Thin Wall. Struct. 50 (2012) 128–140.

[5] R. Sause, B. Kim, M.R. Wimer, Experimental study of tubular flange girders, J Struct. Eng., ASCE 134 (2008) 384–392.

[6] R. Sause, Innovative steel bridge girders with tubular flanges, Struct. Infrastruct. Eng. 11 (2015) 450–465.

[7] F. Gao, H.P. Zhu, D.H. Zhang, Experimental investigation on flexural behavior of concrete-filled pentagonal flange beam under concentrated loading, Thin Wall. Struct. 84 (2014) 214–225.

[8] Y.B. Shao, Y.M. Wang, Experimental study on shear behavior of I-girder with concrete-filled tubular flange and corrugated web, Steel Compos. Struct. 22 (2016) 1465–1486.

[9] Y.B. Shao, Y.M. Wang, Experimental study on static behavior of I-girder with concrete-filled tubular flange and corrugated web under concentrated load at mid-span, Eng. Struct. 130 (2017) 124–141.

[10] Y.M. Wang, Y.B. Shao, Stress analysis of a new steel-concrete composite I-girder, Steel Compos. Struct. 28 (2018) 51–61.

[11] Z.Y. Wang, X.L. Li, W.M. Gai, R.J. Jiang, Q.Y. Wang, Q.Z. Zhao, J.C. Dong, T. Zhang, Shear response of trapezoidal profiled webs in girders with concrete-filled RHS flanges, Eng. Struct. 174 (2018) 212–228.

[12] M.F. Hassanein, O.F. Kharoob, Shear strength and behavior of transversely stiffened tubular flange plate girders, Eng. Struct. 32 (9) (2010) 2617–2630.

[13] M.F. Hassanein, O.F. Kharoob, An extended evaluation for the shear behavior of hollow tubular flange plate girders, Thin-Wall Struct. 56 (2012) 88–102.

[14] H.H. Abbas, R. Sause, R.G. Driver, Analysis of flange transverse bending of corrugated web I-girders under in-plane loads, J. Struct. Eng. 133 (3) (2007) 347–355.

[15] H.H. Abbas, R. Sause, R.G. Driver, Simplified analysis of flange transverse bending of corrugated web I-girders under in-plane moment and shear, Eng. Struct. 29 (2007) 2816–2824.

[16] D.J. Carreira, K.H. Chu, Stress-strain relationship for plain concrete in compression, ACI Proc 82 (1985) 797–804.

[17] M.K. Hassan, Behaviour of hybrid stainless-carbon steel composite beam-column joints, Doctoral dissertation, Western Sydney University, Australia, 2016.

[18] L.H. Han, G.H. Yao, Z. Tao, Performance of concrete-filled thin-walled steel tubes under pure torsion, Thin Wall. Struct. 45 (2007) 24–36.

[19] B. Rabbat, H. Russell, Friction coefficient of steel on concrete or grout, J Struct Eng, ASCE 111 (1985) 505–515.

[20] Z. Tao, B. Uy, F.Y. Liao, L.H. Han, Nonlinear analysis of concrete-filled square stainless steel stub columns under axial compression, J Constr. Steel Res. 67 (2011) 1719–1732.

Recent erection methods

7

7.1 General

As shown earlier in this book, the prestressed concrete box girder with corrugated steel webs (CSWs) is currently a well-known type of composite structure that appeared in France in 1986 [1]. Due to its unique lightweight, superior shear performance, high efficiency of prestressing, and low-carbon environment characteristics, it has been widely and continuously used in Asia; especially in Japan and China [1,2] as explained in detail in the second chaper of this book.

The common construction methods for prestressed concrete box girder bridges with CSWs *mainly* include the bracket construction method, cantilever construction method, and incremental launching method. Among them, the cantilever construction method is widely used in long-span prestressed concrete box girder bridges with CSWs.

In recent years, with the aim of improving the construction efficiency of long-span girder bridges with CSWs, engineers have improved the structural form of the traditional hanging basket and promoted the birth of a new construction method called asynchronous pouring rapid construction (APRC). In this method, the CSWs themselves are used as load-bearing components to directly support the hanging basket and girder segments. Compared to the traditional hanging basket, this new moveable multiple platforms system is more convenient and efficient. The APRC method was first introduced and promoted in bridge construction in Japan. Since 2004, Tsukumi River Bridge, Kinugawa Bridge, Shigaraki Seventh Bridge, and Akabuchigawa Bridge (as shown in Fig. 7.1) have been built in Japan consecutively using this construction method. Recently, China has achieved satisfactory results in applying this new technique in the construction of bridges with CSWs. The APRC method was firstly applied to Tudaohe Bridge in Sichuan Province in 2015 [3,4]. Subsequently, it was used in different types of bridges with CSWs, and longer span bridges with CSWs were built. Currently, there are two representative bridges under construction in China, the short-pylon cable-stayed bridge-Shanxi Yunbao Yellow River Bridge [5] and Ningbo Fenghua River Bridge which is the longest span continuous girder bridge with CSWs in China with the main span of 160 m [6,7].

Previous research mostly focused on the basic mechanical properties of prismatic and non-prismatic beams with CSWs [8–12]. In the literature, a comprehensive introduction to the APRC method is insufficient, despite its use in the construction of many prestressed concrete girder bridges with CSWs (Fig. 7.1). As an advanced construction method, the APRC method not only improves construction efficiency, but also reduces safety risk. Hence, it is a developing trend in modern bridge construction that deserves attention in academics and engineers circles. In this chapter, the structure of the new hanging basket system and the whole construction process of the APRC method are systematically introduced based on the latest practical project in China.

Behavior and Design of Trapezoidally Corrugated Web Girders for Bridge Construction: Recent Advances.
DOI: https://doi.org/10.1016/B978-0-323-88437-2.00007-1

A Shigaraki Seventh Bridge   B Kinugawa Bridge

C Tsukumi River Bridge   D Akabuchigawa bridge

Figure 7.1 Applications of box girder bridges with CSWs built by using the APRC method. *APRC*, asynchronous pouring rapid construction; *CSWs*, corrugated steel webs.

7.2 New hanging basket system

7.2.1 Disadvantages of traditional hanging baskets

It is known that using the traditional hanging baskets results in three main problems for bridge girders [13,14], which are discussed in this subsection. In the traditional hanging basket construction process, cracks easily develop in the bottom concrete flange due to the nonuniform stress distribution near the lifting position. Additionally, the previously poured bottom flange concrete has a short time to reach the required strength before pouring the concrete of the top flange. Moreover, the lifting space of CSW segments is limited by using the traditional diamond hanging basket, which causes some inconvenience to the installation of CSWs.

7.2.2 Proposal of new hanging basket construction technology

At present, the traditional hanging basket construction method is no longer suitable for the rapid construction of long-span girder bridges with CSWs. Therefore, the structural type of the traditional hanging basket was improved by Japanese engineers to reduce the construction time of this type of bridges. The number of the construction platforms for the redesigned hanging basket system was increased to three so that the dislocation construction of the top concrete flange, bottom concrete flange, and CSWs on three adjacent segments can be performed simultaneously. Additionally, in this new hanging basket construction technology, CSWs are used as the main load-bearing components to directly resist the reaction force of the hanging basket system.

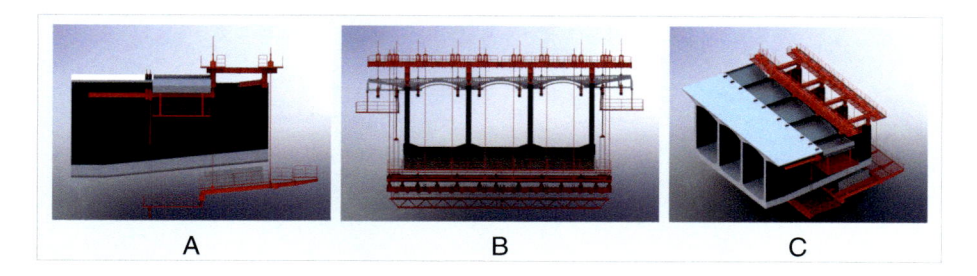

Figure 7.2 New rigid frame hanging basket (A) vertical view; (B) sectional view, and (C) 3D view.

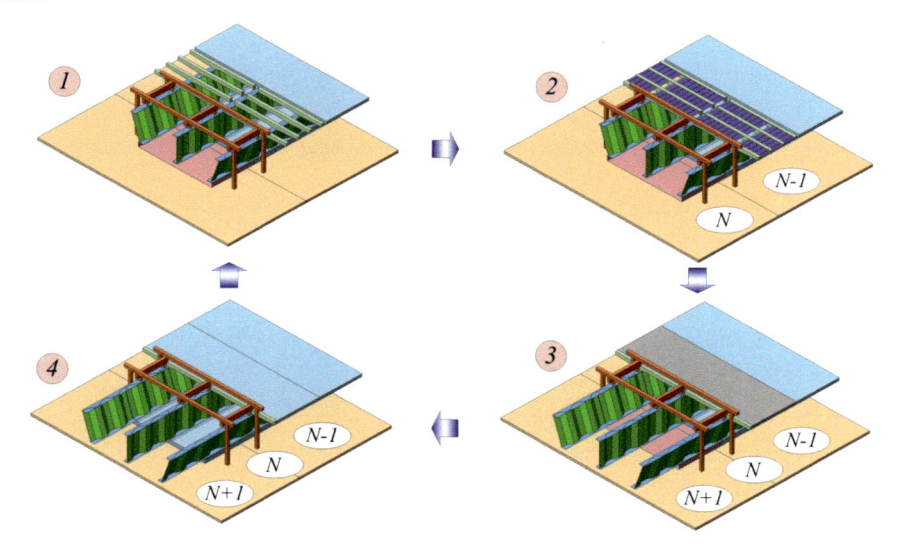

Figure 7.3 Construction steps of box girder bridges with CSWs using APRC. *APRC*, asynchronous pouring rapid construction; *CSWs*, corrugated steel webs.

7.2.3 Structural form of the improved hanging basket system

As shown in Fig. 7.2, the improved hanging basket system is mainly composed of load-bearing truss system, hanging lifting system, formwork system, construction platform, moving system, and anchoring system.

7.3 Asynchronous pouring rapid construction method technical features analysis

7.3.1 Construction steps of standard segments

As shown in Fig. 7.3, the construction steps of the standard segments (i.e., segments between the supported end segments and the middle closure one) of girder bridges with

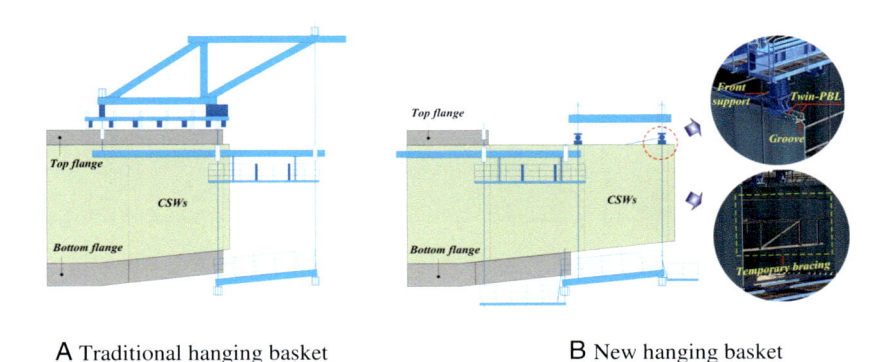

A Traditional hanging basket   **B** New hanging basket

Figure 7.4 Structural types of traditional and new hanging basket systems.

CSWs using the APRC method begin by moving the hanging basket to the N segment, then by:

(1) Temporarily welding the gaps between the top flange plates and bottom flange plates and the perforated plates. Next, the N segmental bottom formwork is installed, while the N-1 segmental top formwork is placed precisely.
(2) Assembling the reinforcement bars of N-1 segmental top flange and N segmental bottom flange simultaneously in parallel with hoisting the N+1 segmental CSWs.
(3) Pouring concretes on the N-1 segmental top flange and N segmental bottom flange simultaneously.
(4) After the compressive strength of the concrete in N-1 segmental top flange reaches the design value, the longitudinal prestressed tendons of N-1 segment are tensioned together with the transverse prestressed tendons of N-2 segmental top flange.
(5) Moving forward the hanging basket to N+1 segment, and proceeding to the next cycle.

7.3.2   Load-bearing corrugated steel webs

The APRC method of girder bridges with CSWs is an improved cantilever construction method that considers the relatively high shear strength of CSWs [15]. As the traditional hanging basket cantilever casting technology has already been very mature (Fig. 7.4A), engineers found that CSWs could be first installed to act as load-bearing components to directly withstand the front support reactions of the hanging basket in the construction of box girder bridges with CSWs. Hence, a new simply supported type of lightweight hanging basket was designed specifically for the APRC method, as shown in Fig. 7.4B. This new hanging basket can move smoothly in the grooves formed by the top steel plate of the CSW and the twin-perfobond leiste (PBL) shear connectors.

During the process of the APRC method, CSWs are typically installed before the top and bottom concrete slabs are poured. Before pouring the top concrete flange, the bottom concrete flange and CSWs together form a stable composite box girder. The thin CSWs, accordingly, bear the weight of the hanging basket system and the weight of the freshly poured concrete. So, the upper parts of CSWs are directly subjected to large concentrated loads. Since the local and global buckling of CSWs are critical problems in construction, transversal temporary bracing is used to enhance the stability of the CSWs, as shown in Fig. 7.4B.

The use of CSWs as load-supporting members skillfully utilizes the high shear strength of the CSWs, making the CSWs not only serve as main components of the bridge itself, but they also provide favorable construction conditions for the APRC method. This greatly reduces the weight of the construction hanging baskets and eliminates many on-site operations such as the erection of formworks and brackets as well as the concrete pouring process associated with the use of traditional concrete webs. Meanwhile, there is enough upper space to hoist and install the CSWs using the new simply supported hanging basket system in the APRC method.

7.3.3 Lightweight hanging basket system

The self-weight of the traditional hanging basket system is quite heavy, which results in significant deformation at the cantilever end. Generally, the weight ratio of the traditional hanging basket to the concrete box girder segment is basically over 0.35 [16]. By using the APRC method, the hanging basket is redesigned to benefit from its lightweight leading to high efficient in erecting long-span girder bridges with CSWs. The structure of the traditional hanging basket and the framework system is greatly simplified because the pre-erected CSWs can provide rigid support for the construction platform. Therefore, the self-weight of the hanging basket and other construction equipment is reduced remarkably. The weight ratio of the new hanging basket to the box girder segment with CSWs may be as low as 0.18 [16]. Hence, the steel consumption of the new hanging basket system becomes about half of that of the traditional hanging basket.

A comparison of the deadweight between the traditional and the new hanging baskets used in Toudao River Bridge (single-box single-chamber) and Fenghua River Bridge (single-box three-chamber) is shown in Fig. 7.5. Note that the deadweight of the new hanging baskets of the APRC technology adopted in Toudao River Bridge and Fenghua River Bridge is about 50t and 120t, respectively [17]. On the opposite, the steel deadweight would probably increase to 145t and 235t, respectively, provided the traditional hanging baskets were planned for these two bridges [17]. Accordingly, for the same bridge, the deadweight of the new hanging basket in the APRC technology is about 34%~51% of that of the traditional hanging basket. This means that a lot of steels were really saved. In general, 187kWh electrical power and 4.4t water are needed and about 2198 kg carbon dioxide are released into the environment to produce a ton of steel. As mentioned above, the weight of the new hanging basket used in the APRC method of Fenghua River Bridge is reduced by 115t. Accordingly, it is clear that the APRC method saves a large amount of electrical power and water, and reduces huge carbon emissions, which creates enormous environmental benefits. Thus, the adoption of the new hanging basket in the construction of girder bridges with CSWs by utilizing the APRC technology can effectively reduce the construction load and the engineering cost, and consequently, the environmental pollution.

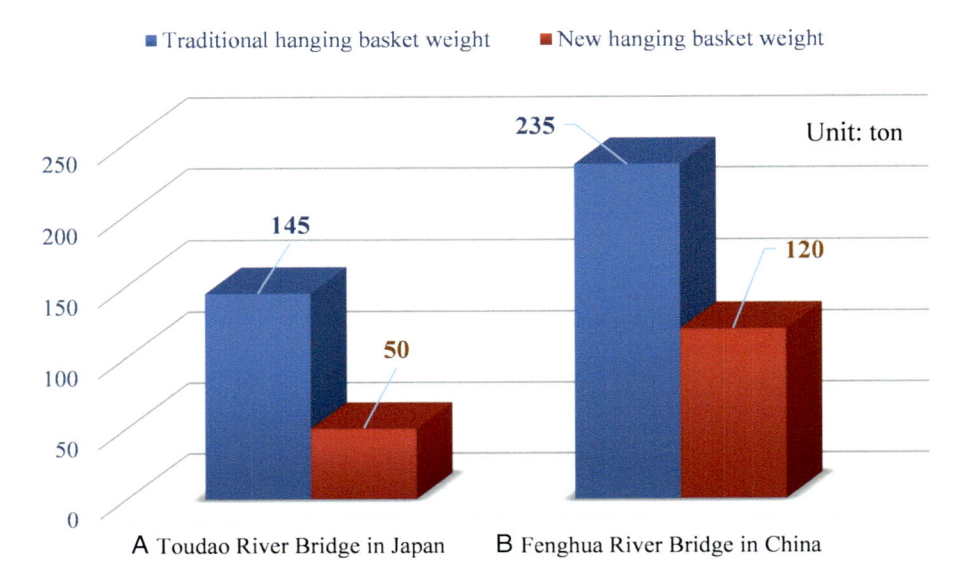

Figure 7.5 Comparison between the weights of new and traditional hanging baskets used in two long-span bridges with CSWs. *CSW*s, corrugated steel webs.

7.3.4 Transformation of hanging basket structural system

As previously mentioned, adopting the traditional diamond hanging basket would increase the steel consumption, construction load and cost, especially for long-span bridges. Under the action of a heavy traditional hanging basket system, the deflection of a long-span box girder with CSWs becomes relatively large and difficult to control during the cantilever construction since the section stiffness of such girder with CSWs is weaker than that of typical concrete box girder. Additionally, the problem of low stability of the box girder with CSWs in cantilever construction stage is very risky by using the traditional hanging baskets, which usually have more mass and higher centre of gravity compared to the girder itself. To ensure the stability and safety of the long cantilever structure with this a heavy basket, a complex post-installed anchorage system is needed, which makes the construction process much more inconvenient.

Given these considerations, a new lightweight simply supported hanging basket system, as shown in Fig. 7.4B, is designed to match the APRC technology based on the traditional cantilevered diamond hanging basket system. This can significantly improve the construction speed of girder bridges with CSWs. In addition, the new light rigid frame simply supported system reduces the possibility of basket overturning and simplifies the previous postinstalled anchorage system. The new hanging basket is supported directly on the CSWs and can move smoothly and safely in the grooves composed of the top flange and the PBL shear connectors with good overall stability.

7.3.5 Extension of working platform

After the traditional diamond hanging basket is installed in place, the construction steps for installing the CSWs, top flange concrete pouring and bottom concrete pouring can only be done step by step in the same segment. Obviously, these three construction steps will affect each other if they are carried out simultaneously. Usually, the pouring of the bottom flange concrete should be performed prior to that of the top flange concrete, so the working platform becomes too limited in space because of the different crossing and overlapping operations.

The prominent problem of mutual interference of operations between top concrete flange and bottom concrete flange in each box girder segment, in addition to the limited working platform associated with the adoption of the traditional diamond hanging basket, is effectively solved by using the APRC method. This new technology expands the box girder segmental working platform from the original single segmental working platform (N) to three adjacent segmental working platforms (N-1, N, N+1), which are (1) N-1 segmental top flange working platform, (2) N segmental bottom flange working platform, and (3) N+1 segmental CSW working platform. Hence, parallel flow construction in the three working platforms can be utilized by this method.

Due to the increase in the number of the working platforms, the efficiency of equipment and workers is remarkably improved. The factory-made segmental CSWs are assembled and connected conveniently in the site, while the top and bottom concrete flanges of the same segment are poured asynchronously. This broadens the prestressing tension working platform, longitudinally expands the effective construction work regions, and rationalizes the construction procedure. All these factors, indeed, promote rapid construction.

7.3.6 Shortening of segmental construction period

Certainly, the traditional cantilever construction technology results in a long construction period and high construction cost, as discussed above. To solve these problems, lightweight CSWs and new hanging baskets are designed to replace thick concrete webs and traditional hanging baskets, respectively, used in cantilever construction.

Compared with the traditional concrete box girders, CSWs significantly reduce the deadweight of the superstructure of the box girder bridges. Usually, the division of box girder segments is determined according to the segmental weight. Hence, the length of box girder segments with CSWs constructed by the APRC method can be increased properly compared to concrete box girder segments. Therefore, the number of divided segments during the cantilever pouring construction stage can be reduced. Additionally, the bottom formwork can move conveniently by utilizing the new handing basket system. Moreover, the APRC method expands the construction platforms making the dislocation construction of the top concrete flange, bottom concrete flange, and CSWs on three adjacent segments to be performed simultaneously. This, definitely, effectively shortens the construction period. Due to its strong adaptation to the environment, the

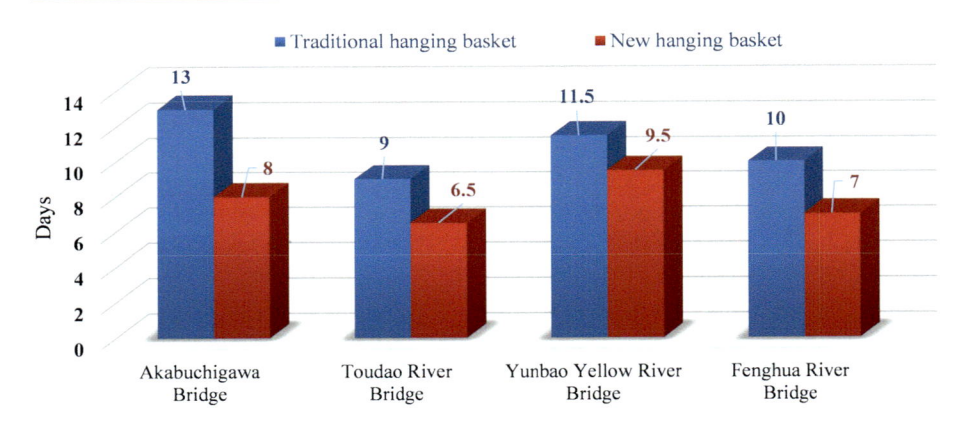

Figure 7.6 Comparison of the segmental construction period with the new and old construction methods for box girder bridges with CSWs. *CSWs*, corrugated steel webs.

APRC method becomes more applicable than the traditional cantilever construction technology for bridge constructions which have a limited schedule in winter.

According to the statistical data of the average segmental construction period of Akabuchigawa Bridge, Toudao River Bridge, Yunbao Yellow River Bridge, and Fenghua River Bridge, shown in Fig. 7.6, it can be realized that the average segmental construction period of the new hanging basket system by considering the APRC technology is three days shorter than that of the traditional basket system. Taking the Toudao River Bridge as an example, which is consists of a single-box and single-chamber cross-section, it can be completed at an average of 6.5 days per segment by utilizing the APRC method, while an average of nine days per segment is achieved under the condition of using the traditional construction method. For this particular case, the construction period of the segment is shortened by 2.5 days; saving 28% of the construction period, especially shortening the pouring period of the top and bottom flanges concrete. Due to the improvement of hanging basket system, the total working days of these four girder bridges with CSWs by using the APRC method can be saved from 44.1% to 56.3% compared to the period taken if the traditional hanging basket construction is used. The total number of the reduced working days for the different projects is represented in Fig. 7.7. Therefore, the APRC method can greatly accelerate the progress of the segmental cantilever construction, shorten the overall construction period and enhance labor efficiency. Accordingly, it provides a good solution for the rapid construction of long-span prestressed girder bridges with CSWs.

7.4   Case study

Fenghua River Bridge, which is a single-box and three-chamber continuous box girder bridge with CSWs, is under construction. Note that it has the largest span and width among the same type of bridges in China. The bridge is crossing the Fenghua River in the east-west direction, and its span is 100+160+100 m. The elevation and

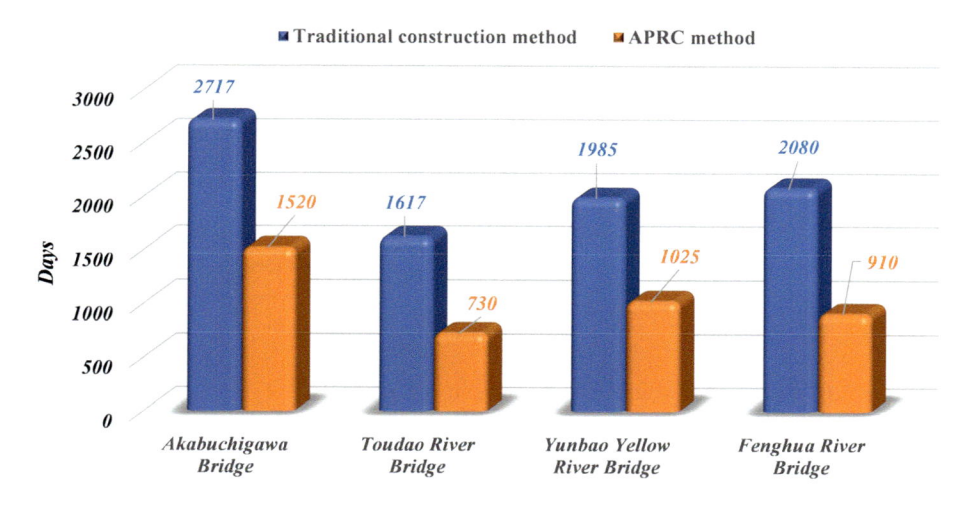

Figure 7.7 Comparison of the total period with the new and old construction methods for box girder bridges with CSWs. *CSWs*, corrugated steel webs.

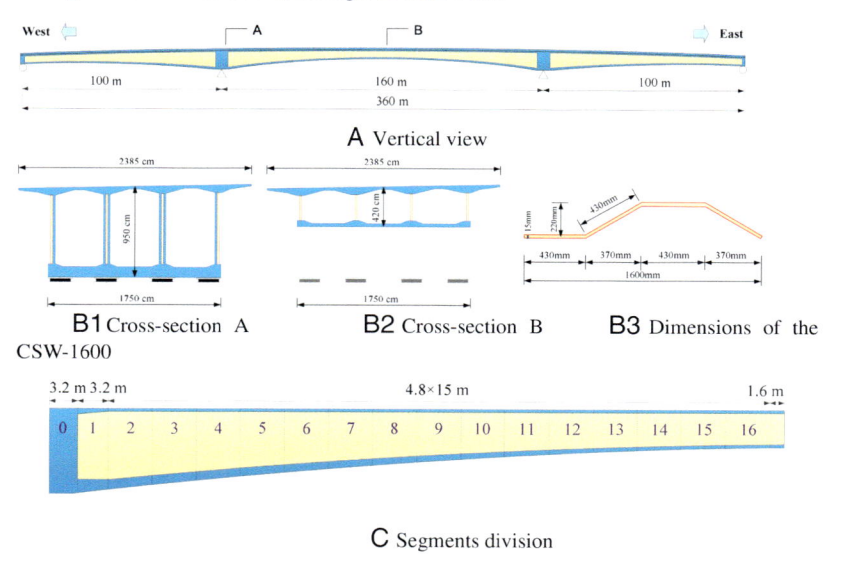

Figure 7.8 Fenghua River Bridge (A) vertical view, (B) cross-sections views, and (C) segments division.

cross-sectional dimensions of the bridge are shown in Fig. 7.8A and B. Cast-in-situ support construction and hanging basket cantilever construction were adopted. Remarkably, most of the segments use the new asynchronous pouring construction technology during the hanging basket cantilever construction.

In order to facilitate the longitudinal connection of the CSWs, and considering that the division of the box girder construction segments should conform with the

Figure 7.9 Bracket construction area of segment #0 and segment #1 in the main span.

Figure 7.10 Nonstandard segment #2 cantilever construction area.

load-carrying capacity of CSWs, the length of the standard longitudinal box girder segment should be taken as an integral multiple of the periodic wavelength of the CSWs (generally it is three times the wavelength). By doing so, the seam could be placed exactly on the flat subpanel of the CSWs. CSW-1600 type, with dimensions shown in Fig. 7.8(b3), is usually adopted in long-span girder bridges with CSWs in China, and this is the case utilized in Fenghua River Bridge. So, the length of the box girder segment of such Bridge is selected as three times the periodic wavelength of CSWs, which is 4.8 m. The divided segments of the maximum cantilever are shown in Fig. 7.8C.

7.4.1   Construction of segment #0 and segment #1

As shown previously, segment #0 and segment #1 (i.e., the construction area shown in Fig. 7.9) were constructed by cast-in-situ brackets. In this method, the supporting steel pipe was erected first. Then, the bracket for the segments was installed and the segments were fixed to the pier temporarily.

When the cast-in-situ segment #0 was completed, the CSW of segment #1 was installed first in position to act as the load-bearing member to support the construction platform. Then, the top and bottom concretes of segment #1 were poured. Additionally, the CSWs of segment #2 were hoisted and connected to prepare for the construction of segment #2 using the APRC method.

7.4.2   Asynchronous construction of non-standard segment #2

After the cast-in-situ segments #0~~#1 were completed, the CSWs of segment #2 were hoisted subsequently to support the hanging basket system. When the hanging basket was installed, the construction of segment #2 bottom flange and the CSWs of segment #3 can be carried out. The construction completion area is shown in Fig. 7.10. When this section was completed, the hanging basket is to be moved forward to start the construction of the following standard segments by the APRC method.

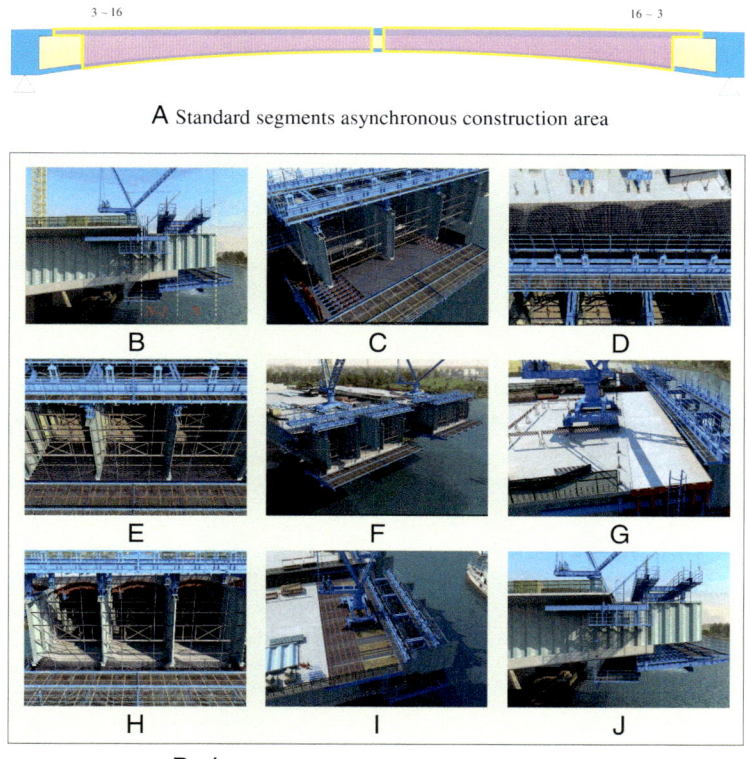

A Standard segments asynchronous construction area

B~J Construction steps of the standard segments

Figure 7.11 Standard segments division and asynchronous construction diagram.

7.4.3 Asynchronous construction of standard segments #3~ #16

Fig. 7.11A shows the standard section area of segments #3 ~ #16. During the process of the asynchronous construction of these segments, some transverse temporary brackets are usually welded between the CSWs to enhance the whole stability and prevent the swinging of the CSWs when the hanging basket moves forward. The main construction steps include moving the hanging baskets forward, erecting the formworks, lashing steel bars, pouring top concrete flange and bottom concrete flange, and tensioning the longitudinal and transverse prestressing reinforcements.

After the construction of the nonstandard segments #0~segemnt #2 are accomplished, the hanging basket is moved forward to segment #3 to construct the subsequent standard segments. Thus, the working platforms will be increased to three, as shown in Fig. 7.11A. The next detailed steps of the APRC method in a standard segment #N, as shown in Fig. 7.11B–J, are to:

(1) Hoist the CSWs of segment #N, and connect the CSWs of segment #N with those of segment #N-1 by high strength bolts and welding, as shown in Fig. 7.11B.

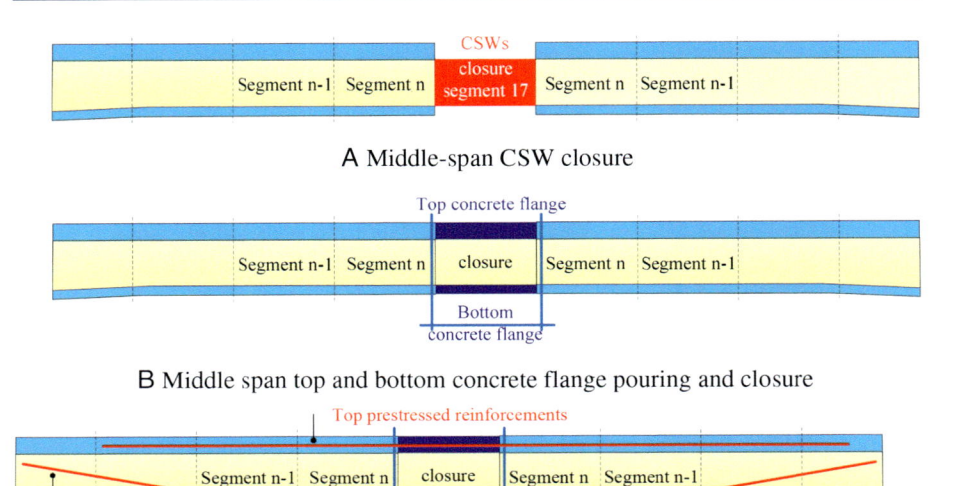

A Middle-span CSW closure

B Middle span top and bottom concrete flange pouring and closure

C Tension the prestressed reinforcements

Figure 7.12 Middle span closure of the main span of Fenghua River Bridge.

(2) Move the hanging basket forward to segment #N and subsequently fix the hanging basket on the CSWs. This is followed by installing the bottom flange formwork of segment #N and the top flange formwork of #N-1 simultaneously, as shown in Fig. 7.11C. Then, assemble the reinforcement bars of the top flange of segment #N-1 and the bottom flange of segment #N, as shown in Fig. 7.11D–E. Meanwhile, hoist the CSWs of segment #N+1, as shown in Fig. 7.11F.

(3) Pour the top flange concrete of segment #N-1 and the bottom flange concrete of segment #N, as shown in Fig. 7.11G and H, then, cure the concrete.

(4) After the concrete strength reaches the design strength, tension the longitudinal prestressed tendons of top flange of segment #N-1 and transverse prestressed tendons of top flange of segment #N-2, as shown in Fig. 7.11I. By doing so, a typical section construction cycle of the APRC method is completed.

(5) Move forward the hanging basket to segment #N+1, and proceed to the next cycle, as shown in Fig. 7.11J.

7.4.4 Construction of closure segment

As shown in Fig. 7.12A, the sides of the CSWs are connected first to ensure the construction safety. The construction processes of the side span closure segment and mid-span closure segment are as follows:

(1) Set up the support to construct side-span cast-in-place segments, then close the side span and apply the tensile stressing of the side-span prestressing reinforcements.

(2) Close the CSWs of the mid-span first (Fig. 7.12A). Then, pour the concrete of the bottom and top flanges in order when the frameworks are in place (Fig. 7.12B). Finally, the mid-span prestressing reinforcements are tensioned when the concrete has reaches its sufficient compression strength (Fig. 7.12C).

(3) Remove the temporary connection of the main piers, and tension the external prestressed reinforcements. Finish follow-up works such as the bridge deck pavement, auxiliary facilities construction, CSWs anti-corrosion coating, etc.

References

[1] W.Q. Deng, M. Zhou, M.F. Hassanein, J.D. Zhang, Liu, L. An, Growth of prestressed concrete bridges with corrugated steel webs in China, P I Civil Eng-Civ En 171 (2) (2017) 77–84.

[2] J. He, S.H. Wang, Y.Q. Liu, L. Zhang, C.X. Li, Mechanical behavior of a partially encased composite girder with corrugated steel web: interaction of shear and bending, Engineering 3 (6) (2017) 806–816.

[3] Z. Lei, H.L. Chen, S. Zhao, The SCC cast-in-cantilever construction method of corrugated steel webs composite girder bridge, Shanxi Arch. 42 (14) (2016) 147–148 Chinese.

[4] Y. Guo, Application of corrugated steel web plate bearing type hanging basket construction technology in the corrugated steel web PC box girder bridge, Constr. Des. Proj. 3 (2) (2017) 117–120 Chinese.

[5] M.J. Jie, Application of dislocation method in construction of prestressed concrete box girder bridge with corrugated steel web, Railway Eng. 4 (2017) 45–47 Chinese.

[6] M. Zhou, Z. Liu, J.D. Zhang, L. An, Z. He, Equivalent computational models and deflection calculation methods of box girders with corrugated steel webs, Eng. Struct. 127 (2016) 615–634.

[7] M. Zhou, J.D. Zhang, J.T. Zhong, Y. Zhao, Shear stress calculation and distribution in variable cross sections of box girders with corrugated steel webs, J. Struct. Eng. 142 (6) (2016) 04016022.

[8] M. Leblouba, S. Barakat, M. Maalej, S. Al-Toubat, A.S. Karzad, Normalized shear strength of trapezoidal corrugated steel webs: improved modeling and uncertainty propagation, Thin Walled Struct. 137 (2019) 68–80.

[9] B.S. Zhang, W.Z. Chen, J. Xu, Mechanical behavior of prefabricated composite box girders with corrugated steel webs under static loads, J. Bridge Eng. 23 (10) (2018) 04018077.

[10] X.C. Chen, M. Pandey, Z.Z. Bai, F.T.K. Au, Long-term behavior of prestressed concrete bridges with corrugated steel webs, J. Bridge Eng. 22 (8) (2017) 04017040.

[11] M.F. Hassanein, A.A. Elkawas, A.M. El Hadidy, M. Elchalakani, Shear analysis and design of high-strength steel corrugated web girders for bridge design, Eng. Struct. 146 (2017) 18–33.

[12] J. Papangelis, N. Trahair, G. Hancock, Direct strength method for shear capacity of beams with corrugated webs, J. Constr. Steel Res. 137 (2017) 152–160.

[13] K.B. Jiang, Y. Ding, H.Y. Gao, W.X. Wang, Y.Z. Zhou, Comparative study on falsework scheme of steel webs for PC composite box-girder with corrugated steel webs during cantilever construction, Ind. Constr 42 (9) (2012) 120–124 Chinese.

[14] Y.S. Lin, Y.B. Zhang, S.Z. Qiang, Suspended basket design of wider bridge and technique of cantilever construction, J. Railway Eng. Soc. 2 (2006) 39–42 Chinese.

[15] A.A. Elkawas, M.F. Hassanein, M.H. El-Boghdadi, Numerical investigation on the non-linear shear behaviour of high-strength steel tapered corrugated web bridge girders, Eng. Struct. 134 (2017) 358–375.

[16] J.Q. Lei, Bridge cantilever construction and design, China Communications Press, Beijing, 2000 Chinese.

[17] S.Q. Li, S. Wang, C.Q. Zhang, Design and manufacture of corrugated steel webs, China Communications Press, Beijing, 2011 Chinese.

Future research

8.1 Recommendations

Based on the recent advances in the behavior and design of trapezoidally corrugated web girders provided in this book for bridge construction, it is recommended that:

1. The realistic support condition at the juncture is nearly fixed for the case of t_f/t_w greater than 3.0. Accordingly, it is recommended to use this real support condition in available interactive shear buckling strength formulas.
2. It is better for designers to take into account the design models given in this book for different cross-sections and until design models are given by different specifications.
3. The optimum design of the corrugated web girders may be reached through following the behavior discussed in this book for different cross-sections.
4. The asynchronous pouring rapid construction (APRC) has been found to raise the construction efficiency compared with traditional hanging basket cantilever construction.

8.2 Trends for future relevant works

Experimental and theoretical investigations should be continued to study the following points (which are of urgent importance):

1. Effect of static and dynamic loads on corrugated web girders made of new stainless steel materials. By doing so, the design will balance the structural integrity of the corrugated web girders with the life-cycle economic and environmental impacts of stainless steel.
2. The influence of static and dynamic loads on corrugated web girders made of ultra-high-strength steels.
3. Impact of static and dynamic loads on corrugated web girder connections.
4. The influence of static and dynamic loads on corrugated web girders with large web opening used for inspections of box-girders, taking into account different materials and cross-sections.
5. Fire safety of the corrugated web girders should be investigated. As a known fact, when metallic structures are exposed to fires, the resistance of their members is greatly reduced. Despite that bridges fires have a low probability compared with conventional steel structures, they are considered nowadays as high-consequence incidents.

Behavior and Design of Trapezoidally Corrugated Web Girders for Bridge Construction: Recent Advances.
DOI: https://doi.org/10.1016/B978-0-323-88437-2.00006-X

Index

Printed in the United States
by Baker & Taylor Publisher Services